Ezra S Winslow

The Foreign and Domestic Commercial Calculator

Ezra S Winslow

The Foreign and Domestic Commercial Calculator

ISBN/EAN: 9783337397326

Printed in Europe, USA, Canada, Australia, Japan

Cover: Foto ©Suzi / pixelio.de

More available books at **www.hansebooks.com**

THE

FOREIGN AND DOMESTIC

COMMERCIAL CALCULATOR;

OR,

A COMPLETE LIBRARY OF NUMERICAL, ARITHMETICAL, AND MATHEMATICAL FACTS, TABLES, DATA, FORMULAS, AND PRACTICAL RULES FOR THE MERCHANT AND MERCANTILE ACCOUNTANT.

BY

E. S. WINSLOW.

Author of "Comprehensive Mathematics," "Computists' Manual," "Machinists' and Mechanics' Practical Calculator and Guide," "Tin-plate and Sheet-Iron Workers' Monitor."

Fourth Edition. Enlarged.

BOSTON:
PUBLISHED BY THE AUTHOR.
1867.

PREFACE
TO THE COMPREHENSIVE MATHEMATICS.

On presenting this work to the public, it may be proper to state that it has been designed and written mainly for the practical man. It contains a vast array of Numerical, Arithmetical, and Mathematical facts, tables, data, formulas, and rules, pertaining to a great variety of subjects, and applicable to a diversity of ends, as well as much information of a more general nature, valuable to the artisan, and commercial classes; thus meeting the wants, in an eminent degree, of the lovers of the exact sciences, and the practical wants of students in the mathematics.

The facts and data alluded to have been gathered, with much care and patience, from a great variety of sources, or derived, often by toilsome investigations, from known and accredited truths. The care that has been taken in respect to these, it is thought, should secure for this particular department reliance and trust.

The tables, which are numerous, have, with few exceptions, been composed and arranged expressly for the work, and a confidence is felt that they may be relied on for accuracy.

From the valuable works of Dr. Ure, Adcock, Gregory, Grier, Brunton; from the publications of the transactions of London, Edinburgh, and Dublin Philosophical Societies; and from the publications by the Smithsonian Institute, much valuable information has been gained, relating mainly to machinery and the arts; and to these sources the author feels indebted.

The conciseness with which the work has been generally written would, perhaps, be found an objection, were it not that all the propositions and problems of intricacy are accompanied with examples and illustrations, and, in the matters of Geometry, additionally accompanied with diagrams. The whole, it is thought, will appear clear to him who consults it. A prominent feature in the design has been to produce a useful work, and one which in the way of price shall be readily accessible to all.

PREFACE

TO THE FOREIGN AND DOMESTIC COMMERCIAL CALCULATOR.

This work is composed of the first four sections of the author's "COMPREHENSIVE MATHEMATICS." It was thought advisable to publish this portion of that work in a separate form on account of price; more especially as it contains all of a commercial nature treated of in that work. Indeed, the contents of that work were arranged expressly to this end. The Table of Contents in both works is the same. The work being stereotyped, this could not well be avoided. The Table of Contents, therefore, in either work, is that of the "COMPREHENSIVE MATHEMATICS," and the first four sections thereof; that is, Section I., Section II., Section III., and Section A., is that of the "Foreign and Domestic Commercial Calculator."

PREFACE

TO THE TIN-PLATE AND SHEET-IRON WORKERS' MONITOR.

This work is composed of Section VI. of the author's "COMPREHENSIVE MATHEMATICS," with portions of other sections of that work. It embraces all that is contained in the last-mentioned work of special interest to the Tinsmith, as such. It may be relied on for accuracy in all particulars, and is believed to be the first and only reliable work of the kind ever published. It is published in separate form on account of price, and with the view of affording apprentices and students every possible facility of obtaining it. It contains over 100 pages, nearly 50 diagrams, and step-by-step directions for constructing, *mechanically*, not less than 30 unlike and different patterns, embracing all of the more difficult and complicated in use, and several of new and beautiful designs.

CONTENTS.

SECTION A.

	PAGE
Foreign Moneys of Account	a 1
Foreign Linear and Surface Measures	a 14
Foreign Weights	a 23
Foreign Liquid Measures	a 35
Foreign Dry Measures	a 44
Custom House Allowances on Dutiable Goods, &c.	a 51
Table of Established Tares	a 52

SECTION I.

MONEYS OF ACCOUNT, COINS, WEIGHTS, AND MEASURES OF THE UNITED STATES; FOREIGN GOLD COINS, &C.

	PAGE
EXPLANATIONS OF SIGNS	12
Moneys of Account of the United States	13
Comparative Value of Gold and Silver	13
Gold, pure; value of, by weight	15
Mint Gold, Standard of, &c.	15
Gold Coins, their weights and values	15
Silver, pure; value of, by weight	16
Mint Silver, Standard of, &c.	16
Silver Coins, their weights and values	16
Copper Coins, &c.	16
Present Par Value of Silver Coins issued prior to June, 1853	17
Currencies of the different States of the Union	17
The Metrical System of Weights and Measures	18
Foreign Gold Coins, TABLES of, &c.	19
Foreign Silver Coins, Values of,	25

WEIGHTS AND MEASURES.

	PAGE
LONG OR LINEAR MEASURE	25
Cloth Measure	25
Land Measure	25
Engineer's Chain	25
Shoemaker's Measure	26
Miscellaneous Measures	26
SQUARE OR SUPERFICIAL MEASURE	26
Measure for Land	26
Circular Measure	27
CUBIC OR SOLID MEASURE	27
GENERAL MEASURE OF WEIGHT,	28
Gross Weight	28
Troy Weight	28
Apothecaries' Weight,	28
Diamonds, Measure of Value, &c.	28
LIQUID MEASURE	28
Imperial Liquid Measure	29
Ale Measure	29
DRY MEASURE	29
Imperial Dry Measure	30

SECTION II.

MISCELLANEOUS FACTS, CALCULATIONS, AND MATHEMATICAL DATA.

	PAGE
SPECIFIC GRAVITIES, TABLES OF,	31
Weight per Bushel of Articles	35
Weight per Barrel of Articles	35
Weights of different Measures of various Articles	35
Weight of Coals, &c., TABLES	35, 55
Practical Approximate Weight in Pounds of Various Articles	36
ROPES AND CABLES	36

CONTENTS.

	PAGE
Weight and Strength of Iron Chains	37
Comparative Weight of Metals, TABLE	38
Weight of Rolled Iron, Square Bar, TABLES	38
Weight of Various Metals, different Forms of Bar	39
Weight of Round-rolled Iron, TABLE	40
Weight of Cast-iron Prisms of different forms, &c	40
Weight of Flat-rolled Iron, TABLE	42
Weight of Different Metals, in Plate	44
THE AMERICAN WIRE GAUGE	45
The Values of the Nos. American Wire Gauge and Birmingham Wire Gauge, in the United States, inch, TABLES of	45
The Number of Linear Feet in a Pound of different kinds of Wire of different Sizes, TABLE of, &c.	46
Characteristics, &c., of Alloys of Copper and Zinc,—BRASS	47
The Weight per Square Foot of different Rolled Metals of different thicknesses by the Wire Gauge, TABLE	48
TIN PLATES, Sizes, &c., TABLE	49
Sheet Iron, Sheet Zinc, Copper Sheathing, Yellow Metal, Weight of, &c	49
Capacity in Gallons of Cylindrical Cans, &c., TABLE	50
Weight of Pipes	52
Weight of Pipes, TABLE	53
Weight of Cast-iron and Lead Balls	54
Weight of Hollow Balls or Shells	54
Analysis of Coals	55
Weight, Heating Power, &c., of Coals and other kinds of Fuel, TABLE	55
MENSURATION OF LUMBER	56
Board Measure	56
To Measure Square Timber	56
To Measure Round Timber	56
TABLE relative to the Measurement of Round Timber	57
To find the Solidity of the greatest Rectangular Stick that can be cut from a Log of Given Dimensions	58
To find the Solidity of the greatest Square Stick that can be cut from a Round Stick of Given Dimensions	59
To find the Contents of a Log in Board Measure	59
GAUGING	60

	PAGE
To find the Dimensions of Vessels of different Forms, for holding Given Quantities	62
CASK GAUGING, all Forms of Casks	63
To find the Contents of a Cask, the same as would be given by the Gauging Rod	66
To find the Diagonal and Length of a Cask	66
ULLAGE	67
To find the Ullage of a Standing Cask	67
To find the Ullage when the Cask is upon its Bilge	67
To find the Quantity of Liquor in a Cask by its Weight	68
Customary Rule by Freighting Merchants for finding the Cubic Measurement of Casks	68
TONNAGE OF VESSELS, to Calculate	69
OF CONDUITS, OR PIPES	70
To find the requisite thickness of a Pipe to support a Given Head of Water	70
To find the Velocity of Water passing through a Pipe	71
To find the Head of Water requisite to a Required Velocity through a Pipe	71
To find the Quantity of Water Discharged by a Pipe in a Given Time	71
To find the Specific Gravity of a Body heavier than Water	72
To find the Specific Gravity of a Body lighter than Water	72
To find the Specific Gravity of a Fluid	72
To find the Quantity of each of the several Metals composing an Alloy	72
To find the Lifting-power of a Balloon	73
To find the Diameter of a Balloon equal to the Raising of a Given Weight	73
To find the Thickness of a Hollow Metallic Globe that shall have a Given Buoyancy in a Given Liquid	73
To Cut a Square Sheet of Metal so as to form a Vessel of the Greatest Capacity the Sheet admits of	73
Comparative Cohesive Forces of Substances, TABLE	74
Alloys having a Tenacity greater than the Sum of their Constituents	74

CONTENTS.

	PAGE
Alloys having a Density greater than the Mean of their Constituents	75
Alloys having a Density less than the Mean of their Constituents	75
Relative Powers of different Metals to Conduct Electricity	75
Dilations of Solids by Heat, TABLE	75
Melting Points of Metals and other Substances, TABLE	76
Relative Powers of Substances to Radiate Heat, TABLE	76
Boiling Points of Fluids	76
Freezing Points of Fluids	77
Expansion of Fluids by Heat	77
Relative Powers of Substances to Conduct Heat	77
Ductility and Maleability of Metals	77
Quantity per cent. of Nutritious Matter contained in different Articles of Food	78
Standard, &c., of Alcohol	78
Quantity per cent. of Absolute Alcohol contained in different Pure Liquors, Wines, &c., TABLE	78
Proof of Spirituous Liquors	78
Comparative Weight of Timber in a Green and Seasoned State, TABLE, &c.	79
Relative Power of different kinds of Fuel to Produce Heat, TABLE	79
Relative Illuminating Power of different Materials, Table and Remarks	80
THERMOMETERS, different kinds, to Reduce one to another, &c.	82
HORSE-POWER	83
Animal Power	83
STEAM, TABLES in relation to, &c.	83, 308
Velocity and Force of Wind, TABLE	84
Curvature of the Earth	84, 213
Degrees of Longitude, Lengths of, &c.	84
TIME, with respect to Longitude	84
Velocity of Sound	84
Velocity of Light	85
GRAVITATION	85, 302
Area of the Earth, its Density, &c.	85
Chemical Elements	86
Elementary Constituents of Bodies, TABLE	87
Combinations by Weight of the Gases in forming Compounds, TABLE	87
Combinations by Volume of the Gases, their Condensation, &c., in forming Compounds	89
Atomic Weight	89

	PAGE
Chemical and other Properties of Various Substances	90

SECTION III.

PRACTICAL ARITHMETIC.

VULGAR FRACTIONS	95
Reduction of Vulgar Fractions	95
Addition of Vulgar Fractions	99
Subtraction of Vulgar Fractions	99
Division of Vulgar Fractions	100
Multiplication of Vulgar Fractions	100
Multiplication and Division of Fractions Combined	101
CANCELLATION	96, 97, 102
To Reduce a Fraction in a higher, to an equivalent in a given lower denomination	102
To Reduce a Fraction in a lower, to an equivalent in a given higher denomination	102
To Reduce a Fraction to Whole Numbers in lower given denominations	103
To Reduce Fractions in lower denominations to given higher denominations	103
To work Vulgar Fractions by the Rule of Three, or Proportion	104
DECIMAL FRACTIONS	104
Addition of Decimals	105
Subtraction of Decimals	105
Multiplication of Decimals	106
Division of Decimals	106
Reduction of Decimals	107
To work Decimals by the Rule of Three	108
Proportion, or Rule of Three	109
Compound Proportion	110
Conjoined Proportion, or Chain Rule	112
PERCENTAGE	114
INTEREST	120
Compound Interest	122
Bank Interest, or Bank Discount	127
DISCOUNT	129
Compound Discount	129
Profit and Loss	130
Equation of Payments	132
General Average	134
Assessment of Taxes	136
Insurance	136
Life Insurance	136
Fellowship	138

viii

CONTENTS.

	PAGE
Alligation	139
Involution	141
Evolution	141
To Extract the Square Root	142
To Extract the Cube Root	143
To Extract any Root	145
Arithmetical Progression	146
Geometrical Progression	150
ANNUITIES	154
Of Installments generally	164
PERMUTATION	166
COMBINATION	167
PROBLEMS	169

SECTION IV.

GEOMETRY.

DEFINITIONS, CONSTRUCTION OF FIGURES, &c.	172
To Bisect a Line	176
To Erect a Perpendicular	176
To Let Fall a Perpendicular	176
To Erect a Perpendicular on the end of a Line	177
To draw a Circle through any three points not in a straight line, and to find the Centre of a Circle, or Arc	177
To find the Length of an Arc of a a Circle approximately by Mechanics	177
From a given Point to draw a Tangent to a Circle	177
To draw from or to the Circumference of a Circle, lines tending to the Centre, when the latter is inaccessible	177
To describe an Oval Arch on a given Conjugate Diameter	178
To describe an Oval of a given Length and Breadth	178
To describe an Arc or Segment of a Circle of Large Radius	179
To describe an Oval Arch, the Span and Rise being given	179
Gothic Arches, to draw	180
Polygons, to construct	181
Polygons, to inscribe in a given Circle	181
Polygons, to circumscribe about a given Circle	181
To produce a Square of the same Area as a given Triangle	181
To construct a Parabola	182, 355
To Construct a Hyperbola	182, 340
To bisect any given Triangle	182

	PAGE
To draw a Triangle equal in Area to two given Triangles	183
To describe a Circle equal in Area to two given Circles	183
To construct a Tothed, or Cog-Wheel	183
OF THE CONIC SECTIONS	184
MENSURATION OF LINES AND SUPERFICIES.	
TRIANGLES	185
Of Right-Angled Triangles	186
Of Oblique-Angled Triangles	187
To find the Area of a Triangle	188
To find the Hypotenuse of a Triangle	189
To find the Base, or Perpendicular, of a Triangle	188, 189
To find the Height of an inaccessible Object	189
To find the Distance of an inaccessible Object	190
To find the Area of a Square, Rectangle, Rhombus, or Rhomboid	190
To find the Area of a Trapezoid	191
To find the Area of a Trapezium	191
OF POLYGONS, TABLE, &c.	194
To find the Perpendicular of a Rhombus, Rhomboid, or Trapezoid	192
To find the Diagonal of a Rhombus, Rhomboid, or Trapezoid	192
To find the Area of a regular or irregular Polygon	195
CIRCLE	196
The Circle and its Sections	197
To find the Diameter, Circumference, and Area of a Circle	198
To find the Length of an Arc of a Circle	199
To find the Area of a Sector of a Circle	201
To find the Area of a Segment of a Circle	201
To find the Area of a Zone	202
To find the Diameter of a Circle of which a given Zone is a part	202
To find the Area of a Crescent	202
To find the Side of a Square that shall contain an Area equal to that of a given Circle	202
To find the Diameter of a Circle that shall have an Area equal to that of a given Square	202
To find the Diameters of three equal circles the greatest that can be inscribed in a given Circle	202

CONTENTS.

	PAGE
To find the Diameters of four equal circles the greatest that can be inscribed in a given Circle	202
To find the Side of a Square inscribed in a given Circle	203
To find the Diameter of a Circle that will circumscribe a given Triangle	203
To find the Diameter of the greatest Circle that can be Inscribed in a given Triangle	203
To divide a Circle into any number of Concentric Circles of equal Areas	204
To find the Area of the space contained between two Concentric Circles	205
ELLIPSE	205
To find the Area of an Ellipse	207
To find the Length of the Circumference of an Ellipse	207
To find the Area of an Elliptic Segment	207
PARABOLA	209
To find the Area of a Parabola	210
To find the Area of a Zone of a Parabola	210
To find the Altitude of a Parabola	210
To find the Length of a Semi-parabola	210
HYPERBOLA	211
To find the Length of a Semi-hyperbola	212
To find the Area of a Hyperbola	212
CYCLOID, and EPICYCLOID	212
To find the Length of the Curve of a Cycloid	213
To find the Area of a Cycloid	213
To find the Distance of Objects at Sea, &c.	213
STEREOMETRY, OR MENSURATION OF SOLIDS.	
OF PRISMS	214
Of Right Prisms or Cubes	215
OF PARALLELOPIPEDONS	215
OF PYRAMIDS	215
OF FRUSTUMS OF PYRAMIDS	216
OF PRISMOIDS	216
OF THE WEDGE	217
OF CYLINDERS	217
To find the Length of a Helix	217
OF CONES	218
OF FRUSTUMS OF CONES	65, 218
OF SPHERES OR GLOBES	219

	PAGE
Of Spherical Segments	219
Of Spherical Zones	220
To find the greatest Cube that can be cut from a given Sphere	220
OF SPHEROIDS	221
Of Segments of Spheroids	221
Of the Middle Frustum of a Spheroid	65, 221
OF SPINDLES	222
Of the Middle Frustum of a Parabolic Spindle	65, 222
OF PARABOLIC CONOIDS	65, 223
OF HYPERBOLOIDS	223
To find the Surface of a Cylindrical Ring	224
To find the Solidity of a Cylindrical Ring	224
OF THE REGULAR BODIES	225
PROMISCUOUS EXAMPLES IN GEOMETRY	226
TRIGONOMETRY	231
TABLES OF SINES, COSINES, TANGENTS, &c.	241
TABLES OF SQUARES, CUBES, SQUARE AND CUBE ROOTS, &c.	245

SECTION V.

MECHANICAL POWERS, MECHANICAL CENTRES, CIRCULAR MOTION, STRENGTH OF MATERIALS; STEAM, THE STEAM ENGINE, ETC.

THE LEVER	271
THE WHEEL AND AXLE	272
THE PULLEY	273
THE INCLINED PLANE	274
THE WEDGE	275
THE SCREW	275
Transverse Strength of Bodies	279
Deflections of Shafts, &c.	286
Resistance of Bodies to Tortion	287
Resistance of Bodies to Compression	289
CENTRES OF SURFACES	291
CENTRES OF SOLIDS	293
CENTRES OF OSCILLATION AND PERCUSSION	294
CENTRE OF GYRATION	298
CENTRAL FORCES	300

CONTENTS.

	PAGE
FLY WHEELS	301
THE GOVERNOR	301
FORCE OF GRAVITY	302
To find the Height of a Stream projected vertically from a Pipe,	303
To find the Power requisite to project a Stream to any given Height	303
OF PENDULUMS	304
SCREW-CUTTING IN A LATHE	305
Table of Change Wheels for Screw-Cutting in a Lathe	308
OF STEAM AND THE STEAM ENGINE	308
Velocity of Projectiles, &c	313
Steam, acting expansively	313
Of the Eccentric in a Steam Engine	314
OF CONTINUOUS CIRCULAR MOTION	314
To find the number of Revolutions made by the last, to one revolution of the first, in a train of Wheels and Pinions	315
The distance from Centre to Centre of two Wheels to work in contact given, and the ratio of Velocity between them, to find their Requisite Diameters	317
To find the Velocity of a Belt	317
To find the Draft on a Machine	317
To find the Revolutions of the Throstle Spindle	318
To find the Twist given to the Yarn by the Throstle	318
TEETH OF WHEELS, &c	318
To construct a Tooth, &c.	319
To find the Horse-Power of a Tooth	319
JOURNALS OF SHAFTS	320
HYDROSTATICS	320
HYDRAULICS	322
WATER-WHEELS	323
To find the Power of a Stream	324
To construct a Water-Wheel to a Given Power and Fall	325
DYNAMICS	326
HYDROSTATIC PRESS	326

SECTION VI.

COVERINGS OF SOLIDS, OR PROBLEMS IN PATTERN CUTTING.

	PAGE
REMARKS AND DEFINITIONS	327
To construct a Pattern for the Lateral Portion of a vessel in the form of a Frustum of a Cone of given diameters and depth	329
To construct a Pattern for the Body of a vessel in the form of a Frustum of a Cone of given dimensions, without plotting the dimensions	332
To construct a Pattern for the Lateral Portion of a Flaring Vessel of given symmetry of outline and given capacity	333
TABLE OF RELATIVE PROPORTIONS, CHORDS, &c.	333
The special tabular figure, the diameter of one end, and the Cubic Capacity of the vessel being given, to find the diameter of the other end	336
To construct a Pattern for the body of a Flaring Vessel of given tabular outline, and given dimensions, without plotting the dimensions	338
The Capacity in gallons of a vessel in the form of a Frustum of a Cone being given, and any two of its dimensions, to find the other dimension	340
To construct Patterns for flaring oval vessels of different eccentricities and given dimensions, Nos. 1, 2, 3	342
To describe the bases for Nos. 1,2,3,	343
OF CYLINDRICAL ELBOWS	348
To construct a Pattern for a Right-angled Cylindrical Elbow	349
To construct Oblique-angled Elbows	352
To construct Right-angled Elliptic Elbows	353
To construct Oblique-angled Elliptic Elbows	353
To construct Right Semi-hyperbolas by intersecting lines	349, 353
To construct the Quadrant of a Circle by intersecting lines	354
To construct the Quadrant of a given Ellipse by intersecting lines	354
To construct the Quadrant of a Cycloidal Ellipse by intersecting lines	354
To describe an Ellipse of given dimensions by means of two Posts, a Pencil, and a String	354
To find the length of the circumference of a given Ellipse	207, 355
To construct a Semi-parabola by intersecting lines	355

CONTENTS.

	PAGE
OVALS, to describe	178, 343, 346, 347
OF CIRCULAR ELBOWS	355
TABLE applicable to Circular Elbows	356
To construct a Right-angled Circular Elbow of 3, 4, 5, 6, 7, or 8 pieces, &c.	355
To construct a Collar for a Cylindrical Pipe of the same diameter as the receiving pipe	359
To construct a Cylindrical Collar of a given Diameter to fit a Receiving-pipe of a greater given Diameter	360
To construct a Cylindrical Collar to fit an Elliptic-cylinder at either right section of the Ellipse	361
To construct a Cylindrical Collar of a given Diameter, to fit a Cylinder of the same Diameter, at any given Angle to the side of the Cylinder	361
To construct a Cylindrical Collar, or Spout, of a given Diameter, to fit a Cylinder of a greater given Diameter, at a given Angle to the side of the Cylinder	362
OF SPOUTS FOR VESSELS	363
Of Pitched or Bevelled Covers	364
To construct a Bevelled Circular Cover of a given Rise and given Diameter	364
To construct a Pattern for a Bevelled Elliptical Cover of a given Rise to fit an Elliptic Boiler of given Diameters	365
To construct a Bevelled Cover of a given Rise, to fit a False-Oval Boiler of given length and width	365
OF CAN-TOPS	366
To construct a Can-top of a given Depth and given Diameters	366
To construct a Can-top of a given Pitch, and given Diameters	367
OF LIPS FOR MEASURES	368
To construct a Lip for a Measure, the Diameter of the Top of the Measure being given	369
OF SHEET PANS	369
To cut the Corners for a Perpendicular-sided Sheet Pan	370
To cut the Corners for an Oblique-sided Sheet Pan	370
To construct a Heart, or Heart-shaped Cake-Cutter	370
To construct a Mouth-piece for a Speaking-Tube	370
To construct a Pattern for the Body of a Circular-bottomed Flaring Coal-Hod, all the curves to be arcs of circles	371
SOLDERS, ALLOYS, AND COMPOSITIONS	373

DEFINITIONS

OF THE SIGNS USED IN THE FOLLOWING WORK.

$=$ *Equal to.* The sign of equality; as 16 oz. $=$ 1 lb.
$+$ *Plus*, or *More*. The sign of addition; as $8 + 12 = 20$.
$-$ *Minus*, or *Less*. The sign of subtraction; as $12 - 8 = 4$.
$\times$ *Multiplied by.* The sign of multiplication; as $12 \times 8 = 96$.
$\div$ *Divided by.* The sign of division; as $12 \div 4 = 3$.
$\sim$ *Difference between the given numbers or quantities;* thus, $12 \sim 8$, or $8 \sim 12$, shows that the less number is to be subtracted from the greater, and the difference, or remainder, only, is to be used; so, too, *height* $\sim$ *breadth*, shows that the difference between the height and breadth is to be taken.
$: :: :$ *Proportion;* as $2 : 4 :: 3 : 6$; that is, as 2 *is to* 4, *so is* 3 *to* 6.
$\sqrt{}$ *Sign of the square root;* prefixed to any number indicates that the square root of that number is to be taken, or employed; as $\sqrt{64} = 8$.
$\sqrt[3]{}$ *Sign of the cube root;* and indicates that the cube root of the number to which it is prefixed is to be employed, instead of the number itself; as $\sqrt[3]{64} = 4$.
² *To be squared, or the square of;* shows that the square of the number to which it is affixed is the quantity to be employed; as $12^2 \div 6 = 24$; that is, that the square of 12, or $144 \div 6 = 24$.
³ Indicates that the cube of the number to which it is subjoined is to be used; as $4^3 = 64$.
. *Decimal point*, or *separatrix*. See DECIMAL FRACTIONS.
—— *Vinculum.* Signifies that the two or more quantities over which it is drawn, are to be taken collectively, or as forming one quantity; thus, $\overline{4 + 6} \times 4 = 40$; whereas, without the vinculum, $4 + 6 \times 4 = 28$; also, $12 - \overline{2 \times 3 + 4} = 2$; and $\sqrt{5^2 - 3^2} = 4$. So, also, $\sqrt{(5^2 - 3^2)} = 4$, and $(4 + 6) \times 4 = 40$.

$\dfrac{4^2}{2}$ $\left\{\begin{array}{l}\text{half of } 4^2 \text{ or} \\ \text{half of the square of 4}\end{array}\right\} = 8.$

$\left(\dfrac{4^2}{2}\right)^2$ (the square of half the square of 4) $= 64$.

$\tfrac{1}{2}b^2$ or $\tfrac{1}{2}(b)^2$ (half the square of b.)
$(\tfrac{1}{2}b)^2$ (the square of half b)
$(2b)^2$ (the square of twice b.)

SECTION I.

MONEYS, WEIGHTS AND MEASURES,

OF THE UNITED STATES; — THEIR DENOMINATIONS, VALUES, COMPARATIVE VALUES, MAGNITUDES, &c.

MONEYS OF ACCOUNT OF THE UNITED STATES.

These are the *mill*, the *cent*, the *dime*, and the *dollar*.

10 mills = 1 cent, 10 cents = 1 dime, 10 dimes = 1 dollar.

The *dollar* is the *unit* or ultimate money of account of the United States, or of what is sometimes called *Federal money*.

In practice, the dime, as a denomination of value, is rejected. Thus,

10 mills = 1 cent, and 100 cents = 1 dollar.

This mark, $, is equivalent to the word *dollar*, or *dollars*, in this money.

COINS OF THE UNITED STATES.

Until June, 1834, the government of the United States estimated gold in comparison with silver as 15 to 1, and in comparison with copper as 850 to 1.

From June, 1834, until February, 1853, the same government estimated gold in comparison with silver as 16 to 1, and in comparison with copper as 720 to 1.

For all time since February, 1853, this government has estimated gold in comparison with silver as $14\frac{7}{8}$ to 1, and in comparison with copper as 720 to 1.

The standard for mint gold with this government until 1834, was 11 parts pure gold and 1 part alloy, the alloy to consist of silver and copper mixed, not exceeding one half copper.

The gold coins, therefore, struck at the United States mint prior to 1834, are 22 carats fine.

14   CURRENCY OF THE UNITED STATES.

In what, until 1834, constituted a dollar of gold coin of United States mintage, there were put 24.75 grains of pure gold; and 27 grains of the standard mint gold of that day were at that time worth $1. Twenty-seven grains of that gold, or gold of that standard, are now, by the present government standard of valuation, worth $1.0652.

The standard for mint silver with this government until 1834, was 1485 parts pure silver and 179 parts pure copper, $= 8\frac{59}{179}$ parts pure silver and 1 part pure copper.

The silver coins, therefore, struck at the United States mint prior to 1834, are $10\frac{231}{416}$ ounces fine.

In that which, until 1834, constituted a dollar of silver coin of this government's mintage, there were put $371\frac{1}{4}$ grains of pure silver; and 416 grains of the standard mint silver of that day were at that time of the value of $1. Four hundred and sixteen grains of that silver, or silver of that standard, are now, by the present government standard of valuation, worth $1.0744.

The *cent*, until 1834, was of pure copper, and weighed 208 grains; since 1834, pure copper, weight 168 grains.

The standard for mint gold with this government is now, and for all time since June, 1834, has been, 9 parts pure gold and one part alloy, the alloy to consist of silver and copper mixed, not exceeding one half silver.

The gold coins, therefore, struck at the United States mint and dated subsequent to 1834, are $21\frac{3}{5}$ carats fine.

The standard weight for these coins is $25\frac{4}{5}$ grains to the *dollar;* and in every $25\frac{4}{5}$ grains of these coins there are $23\frac{22}{100}$ grains of pure gold.

The standard for mint silver with this government is now, and for all time since June, 1834, has been, 9 parts pure silver and 1 part pure copper.

The silver coins, therefore, struck at the United States mint and dated subsequent to 1834, are $10\frac{4}{5}$ ounces fine.

In what, from June, 1834, until February, 1853, constituted a dollar of silver coin of this government's mintage, there were put $371\frac{1}{4}$ grains of pure silver; and $412\frac{1}{2}$ grains of the standard mint silver of that day (the present standard) were worth, from June, 1834, until February, 1853, $1. Four hundred twelve and one half grains of this standard of silver are now worth, by the present standard of valuation, $1.0742.

The standard weight for silver coins with this government at present is 384 grains to the *dollar*.

The new *cent*, established by the Congress of 1856, is 7 parts copper and 1 part nickel, and its legal weight is 72 grains.

The foregoing is not applicable to the *three-cent pieces* of United

CURRENCY OF THE UNITED STATES. 15

States mintage. These pieces were ordered by the Congress of 1850–1851, and an especial standard of purity was assigned them, viz., three parts silver and one part copper; their weight was fixed at 12⅜ grains each, and their current value at *three* cents each. The law of 1853, regulating the currency, does not apply to these. They are now, as in 1851, legally the same. These pieces are worth, even now, less than their nominal values, compared with the present standard of purity and weight for other United States coins. They are worth, by this comparison, 2.863 cents each.

In the preceding calculations, the alloy for gold, in each instance, was taken to consist of equal parts of silver and copper. The law, until 1834, provided that it should consist of 'silver and copper mixed, not exceeding one half copper;' and the present law provides that it shall consist of 'silver and copper mixed, not exceeding one half silver.'

The metals used as alloys were taken at their values as money.

Federal money was established by the Congress of the United States, in 1786.

Boston, June, 1866.

GOLD, — PURE.

24 carats fine = Pure Gold.
1 grain = $0.0429.
23.30859 " = $1.00.
1 dwt. = $1.02966.
1 ounce = $20.5932.

MINT GOLD. — U. S.

Alloy half each, silver and copper.

Nine parts pure gold and one part alloy; or,

21⅗ carats fine = Standard Coin.
1 grain = $0 03876.
25⅘ " = $1.00.
1 dwt. = $0.93023.
1 ounce = $18.60465.

GOLD COINS.— U. S.

	Weight in Grains.	Standard Value.
Double Eagle, - - - -	516	$20.00
Eagle, - - - - - -	258	10.00
Half Eagle, - - - - -	129	5.00
Quarter Eagle, - - - -	64½	2.50
Gold Dollar, - - - - -	25⅘	1.00
Triple Gold Dollar, - - -	77⅖	3.00
Eagle, prior to 1834, ($10½,) - -	270	10.64
Half do., " " " ($5¼,) - -	135	5.32

Private and Uncurrent.	Weight in Grains.	Sales.
A. Bechtler, N. C., $5 piece, - -		$4.75
" " 2½ " - - - -		2.37
" " 1 " - - - -		.93
T. Reed, Georgia, 5 " - - - -		4.75
" " 2½ " - - - -		2.37
" " 1 " - - - -		.93
Moffat, California, 5 " - - - -	129	5.00

SILVER,— PURE.

12 ounces fine = Pure Silver.
1 dwt. = $0.06928.
$346\frac{38}{75}$ grains = $1.
1 ounce = $1.3857.

MINT SILVER.— U. S.

Alloy, all copper.

Nine parts pure silver and one part alloy; or,
10 oz. 16 dwts. fine = Standard Coin.
1 dwt. = $0.062.
384 grains = $1.00.
1 ounce = $1.23958.

SILVER COINS.— U. S.

	Weight in Grains.	Standard Value.
Dollar, - - - - - - - -	384	$1.00
Half Dollar, - - - - - -	192	.50
Quarter Dollar, - - - - - -	96	.25
Dime, - - - - - - -	$38\frac{2}{5}$	.10
Half Dime, - - - - - - -	$19\frac{1}{5}$	.05
Three-Cent Piece, ¾ silver and ¼ copper,	$12\frac{3}{8}$	.03

The copper coins of the United States are the CENT and HALF CENT; they are of pure copper. The weight of the former is 168 grains, and that of the latter, 84 grains.

NOTE.— The silver coins of the United States, issued since February, 1853, are not legal tender in the United States in sums exceeding *five dollars*.

TABLE,

Exhibiting the standard weight and present par value of the silver coins of the United States, of dates subsequent to 1834, and prior to 1853.

	Weight in Grains.	Present par value.
Dollar, - - - - -	$412\frac{1}{2}$	$1.0742
Half Dollar, - . - -	$206\frac{1}{4}$	.5371
Quarter Dollar, - - - -	$103\frac{1}{8}$	.2685
Dime, - - - - -	$41\frac{1}{4}$	.1074
Half Dime, - - - -	$20\frac{5}{8}$	.0537
Three-Cent Piece, - - -	$12\frac{3}{8}$	.03

CURRENCIES OF THE DIFFERENT STATES OF THE UNION.

4 Farthings = 1 Penny, 12 Pence = 1 Shilling, 20 Shillings = 1 Pound.

In Massachusetts, Connecticut, Rhode Island, New Hampshire, Vermont, Maine, Kentucky, Indiana, Illinois, Missouri, Virginia, Tennessee, Mississippi, Texas and Florida, 6 shillings = 1 dollar; $1 = \frac{3}{10}$ £.

In New York, Ohio and Michigan, 8 shillings = 1 dollar; $1 = \frac{2}{5}$ £.

In New Jersey, Pennsylvania, Delaware and Maryland, 7 shillings and 6 pence = 1 dollar; 1 dollar = $\frac{3}{8}$ £.

In North Carolina, 10 shillings = 1 dollar; $1 = \frac{1}{2}$ £.

In South Carolina and Georgia, 4 shillings and 8 pence = 1 dollar; $1 = \frac{7}{30}$ £.

NOTE.— These *currencies*, so called, are nominal at present in a great measure. The denominations serve in the different States as verbal expressions of value. But they are neither the names of the moneys of account in any of the States, nor are they the national names of any of the real moneys in circulation. All values in money in the United States are legally expressed in *dollars*, *cents*, and *mills*.

THE METRICAL SYSTEM OF WEIGHTS AND MEASURES.

In this system, the METRE is the basis, and is one forty-millionth of the polar circumference of the earth.

The METRE is the principal unit measure of length; the ARE of *surface;* the STERE of *solidity;* the LITRE of *capacity;* and the GRAM of *weight.*

The gram is the weight, in a vacuum, of one cubic centimetre of pure water at its maximum density.

The Metre, almost exactly . = 39.3685 U. S. inches.
The Are (100 square metres) = 3.95337 " square rods.
The Stere (a cubic metre) . = 35.31042 " cubic feet.
The Litre (a cubic decimetre) = { 61.0164 " " inches.
                                 { 1.05656 " wine quarts.
The Gram . . . . = 15.44242 " grains.

The divisions by 10, 100, 1,000, of each of these units, are expressed by the same prefixes, viz., *deci, centi, milli;* and the multiples by 10, 100, 1,000, 10,000, of each, by *deca, hecto, kilo, myria.* The former series were derived from the Latin language, the latter from the Greek.

To illustrate with the metre: —

10 *milli*metres = 1 *centi*metre, 10 centimetres = 1 *deci*metre, 10 decimetres = 1 METRE, 10 METRES = 1 *deca*metre, 10 decametres = 1 *hecto*metre, 10 hectometres = 1 *kilo*metre, 10 kilometres = 1 *myria*metre.

In commerce, the ordinary weight is the kilogram, and 100 kilograms (usually called kilos) = 1 quintal; 10 quintals = 1 millier, or tonneau. The kilogram = 15,442.42 ÷ 7000 = 2.20606 avoirdupois pounds.

In practice, the terms *milliare, deciare, decare, kiloare,* and *myriare* are usually dropped, and

100 centare = 1 are; 100 ares = 1 hectare.

Also the terms *millistere, hectostere, kilostere,* and *myriastere,* are usually rejected, and 100 centisteres = 1 decistere; 10 decisteres = 1 stere; 10 steres = 1 decastere = 353.1042 cubic feet.

1 centiare (square metre) = 1.19589413 square yards.
1 kilometre . . . = 0.62135 statute miles.
1 hectare . . . = 2.471 = U. S. acres.
1 kilolitre . . . = 1 stere = 61,016.403233 cubic in.
A hectolitre = 26.41403 wine gallons = 2.83741 Winchester bush.

NOTE.—The system is the one recommended by the Statistical Congress of 1865 as a general system of weights and measures to be adopted by all nations.

FOREIGN GOLD COINS.

Note.— The coins of any country, both gold and silver, circulating as foreign in any other, particularly those of the smaller denominations, are usually held at an estimate below their *standard* par value, compared with the money standard of the country in which they circulate as foreign. Many of them, more particularly the silver, having circulation in the United States, are much worn and otherwise depreciated. In some instances, owing to frequent changes made both with regard to weight and purity, certain of them, having the same name and general appearance, bear a premium at home; others, a discount. Others, again, can hardly be said to have a definable value anywhere. The par value of the old pistole of Geneva, for instance, weighing 103¼ grains, is $3.985, while that of the new, weighing 87¾ grains, would, at the same degree of purity, be worth but $3.386 ; whereas, owing to its higher standard of fineness, its par value is $3.443. The ducat of Austria, coined in 1831, weighs 53¼ grains, — its purity is 23.64, and its par value $2.269; while the half sovereign, closely resembling the ducat, coined in 1835, and weighing 87 grains, has a purity only of 21.64, and a par value, consequently, of but $3.378. The *circulating* value of the ducat in the United States, in general, is $2.20, and that of the half sovereign of Austria, $3.25.

ARGENTINE REPUBLIC.	Standard of purity in carats.	Standard weight in grains.	Par value in Federal money.	Circulating value in Federal money.	Par value per grain. cts.
Doubloon to 1832,	19.56	418	$14.671	$	3.50
" to "	20.83	415	15.512		3.73
AUSTRIA.					
Sovereign, half in proportion, to 1785,	22.00	170	6.711	6.50	3.94
Sovereign, half in proportion, since 1785,	21.64	174	6.756	6.50	3.88
Ducat, double in proportion,	23.64	53½	2.269	2.20	4.24
BELGIUM.					
Sovereign, half in pro.,	22.00	170	6.711		3.94

FOREIGN GOLD COINS.

	Standard of purity in carats.	Standard weight in grains.	Par value in Federal money.	Circulating value in Federal money.	Par value per grain. cts.
Twenty Franc, more in pro.	21.50	99½	$3.840	$3.83	3.85
Ducat,				2.20	
BOLIVIA, COLOMBIA, CHILI, ECUADOR, PERU, NEW GRENADA, and MEXICO. Received by U. S. Government, — those of not less than 20.86 carats fine, at 89 9/10 cts. per dwt.					
Doubloon, (8 E)	20.86	417	15.620	15.60	3.74
Half do.	"	208½	7.810	7.50	"
Quarter do.	"	104¼	3.905	3.75	"
Eighth do.	"	52	1.952	1.75	"
Sixteenth do.	"	26	.976	.90	"
Pistole, half in pro.,				3.75	•
BRAZIL. Received by the U. S. Government, — those of not less than 22 carats fine, at 94 6/10 cts. per dwt.					
Dobraon,	22.00	828	32.719	32.00	3.95
Dobra,	"	438	17.306	17.00	"
Joannes, (*standard variable*)	"	432	17.064	$13 to $17	"
Half do. do. do.	"	216	8.532	$6 to 8.50	"
Moidore, (BBBB) half in pro., (*standard variable*)	21.79	165	6.451	6.00	3.90
Crusado, do. do.	"	16¼	.635		"
DENMARK.					
Christian d'or	21.74	103	4.018		3.90
Ducat, species,	23.48	53⅜	2.254	2.20	4.21
" current,	21.03	48	1.811		3.77
FRANCE. (*Alloy mostly silver.*) Rec'd by U. S. Government, — those the purity of which is not less than 21 6/10 carats fine, at 93 1/10 cts. per dwt.					
Chr. d'or, double in pro.,	21.60	101	3.914	3.90	3.87

	Standard of purity in carats.	Standard weight in grains.	Par value in Federal money.	Circulating value in Federal money.	Par value per grain. cts.
Franc d'or, double in pro.,	21.60	101	$3.914	$3.90	3.87
Louis d'or, " " " to 1786,	21.49	125¼	4.840		3.85
Louis d'or, double in pro., since 1786,	21.68	118	4.573	4.50	3.87
Napoleon (20 F.) double &c.	21.60	99½	3.856	3.83	"
GERMANY.					
BADEN.					
Zehn Gulden, 5 in pro.,	21.60	105½	4.088	4.00	3.87
BAVARIA.					
Carolin,	18.49	149¼	4.952		3.32
Ducat, double in pro.,	23.58	53¾	2.275	2.20	4.23
Maximilian,	18.49	100	3.317		3.31
BRUNSWICK.					
Ducat,	23.22	53½	2.220		4.16
Pistole, double in pro.,	21.60	117¼	4.548		3.87
Ten Thaler, 5 in pro., to 1813,	21.55	202	7.811	7.80	3.86
Ten Thaler, less in pro., since 1813,	21.50	204	7.873	7.80	3.85
HANOVER.					
Ducat, double in pro.,	23.83	53½	2.287	2.20	4.27
George d'or, " " "	21.67	102½	3.987		3.88
Zehn Thaler, 5 " "	21.36	204½	7.838	7.80	3.83
HESSE.					
Ten Thaler, 5 in pro., to 1785,	21.36	202	7.742		"
Ten Thaler, 5 in pro., since 1785,	21.41	203	7.799		3.84
SAXONY.					
Ducat,	23.49	53¼	2.256	2.20	4.21
Augustus d'or, double in pro., since 1784.	21.55	102½	3.964		3.86
WURTEMBURG.					
Carolin,	18.51	147½	4.899		3.32
Ducat,	23.28	53¼	2.235		4.17

FOREIGN GOLD COINS.

	Standard of purity in carats.	Standard weight in grains.	Par value in Federal money.	Circulating value in Federal money.	Par value per grain. cts.
GREAT BRITAIN. (*Alloy, since* 1826, *all copper.*) Rec'd by U. S. Government,—those of 22 carats fine, at 94$\frac{9}{10}$ cts. per dwt.					
Guinea, half in pro., to 1785,	22.00	127	$5.016		3.95
Guinea, half in pro., since 1785,	"	129½	5.111	$5.00	"
Sovereign, half in pro.,	"	123¼	4.866	4.83	"
Five do.	"	616¼	24.332	24.20	"
Sovereign, (*dragon*) half in pro.,	"	122½	4.838	4.80	"
Double Sovereign (*dragon*)	"	246	9.717	9.67	"
GREECE.					
Twenty Drachm, more in pro.,	21.60	89	3.449	3.30	3.87
HOLLAND.					
Ducat,	23.58	53¼	2.263	2.20	4.23
Ryder,	22.00	153	6.043		3.95
Double do.	"	309	12.205		"
Ten Gulden, 5 in pro.,	21.60	104	4.025	4.00	3.87
INDIA.					
Pagoda, star,	19.00	52¾	1.798		3.40
Mohur, (E. I. Co.) 1835.	22.00	180	7.106	6.75	3.95
Half Sovereign, do.				2.41	
BOMBAY.					
Rupee,	22.09	179	7.095		3.96
MADRAS.					
Rupee,	22.00	180	7.106		3.95
ITALY.					
ETURIA, Ruspone,	23.97	161¼	6.935		4.30
GENOA, Sequin,	23.86	53½	2.291		4.28
MILAN, Pistole,	21.76	97½	3.807		3.90
" Sequin,	23.76	53½	2.281		4.26

FOREIGN GOLD COINS.

	Standard of purity in carats.	Standard weight in grains.	Par value in Federal money.	Circulating value in Federal money.	Par value per grain. cts.
MILAN, Twenty Lire, more in proportion,	21.58	99½	$3.853	$3.83	3.86
NAPLES, Ducat, multiples in pro.,	21.43	22½	.865		3.84
NAPLES, Oncetta,	23.88	58	2.485		4.28
PARMA, Doppia, to 1786,	21.24	110	4.192		3.81
" Pistole, since 1796,	20.95	110	4.135		3.75
" Twenty Lire,	21.62	99½	3.860	3.83	3.87
PIEDMONT, Carlino, half in pro., since 1785,	21.69	702	27.321		3.89
PIEDMONT, Pistole, half in pro., since 1785,	21.54	140	5.411		3.86
PIEDMONT, Sequin, half in pro., since 1785,	23.64	53¾	2.280		4.23
PIEDMONT, Twenty Lire, more in pro.,	20.00	99¼	3.563	3.50	3.59
ROME, Ten Scudi, 5 in pro.	21.60	267½	10.368		3.87
" Sequin, since 1760,	23.90	52½	2.251		4.28
SARDINIA, Carlino, ¼ in pro.,	21.31	247½	9.465		3.82
TUSCANY, Zechino, double in pro.,	23.86	53¾	2.302		4.30
VENICE, Zechino, double in pro.,	23.84	·54	2.310		
MALTA.					
Sequin,	23.70	53¼	2.275		4.25
Louis d'or, double and demi in pro.,	20.25	128	4.651		3.63
NETHERLANDS.					
Ducat,	23.52	53¼	2.257		4.21
Zehn Gulden, 5 in pro.,	21.55	103¾	4.013	4.00	3.86
POLAND.					
Ducat,	23.58	53½	2.264		4.23
PORTUGAL.					
Rec'd by the U. S. Government, — those the purity of which is not less than 22 carats fine, at 94 8/10 cts. per dwt.					

	Standard of purity in carats.	Standard weight in grains.	Par value in Federal money.	Circulating value in Federal money.	Par value per grain. cts.
Dobraon, 24,000 reis,	22.00	828	$32.706	$32.00	3.95
Dobra,	"	438	17.301	17.00	"
Joannes, (*standard variable*)	"	432	17.064	$13 to $17	"
Half " " "	"	216	8.532	$6 to 8.50	"
Moidore, 4000 reis, "				$4¾ to $4¼	
Coroa, 5000 "	"	147½	5.83	5.75	"
Milrea,	22.00	19¾	.780		3.95
PRUSSIA.					
Ducat,	23.49	53½	2.255	2.20	4.21
Frederick d'or, double in pro.,	21.60	102½	3.973		3.87
RUSSIA.					
Ducat,	23.64	54	2.291		4.24
Imperial, (10 R.) half in pro., 1801,	23.55	185¼	7.828		4.22
Imperial, (10 R.) half in pro., since 1818,	22.00	199	7.856	7.80	3.95
SICILY.					
Oncia, double in pro.,	20.39	68¼	2.495		3.64
Twenty Lire, more in pro.,	21.60	99½	3.856	3.83	3.87
SPAIN.					
Rec'd by U. S. Government, — those the standard purity of which is not less than 20.86 carats fine, at 89 9/10 cts. per dwt.					
Doubloon (8 S) parts in pro.	21.45	416½	16.031	16.00	3.84
" (8 E) parts as Bolivian, &c.	20.86	417	15.620	15.60	3.74
Pistole, to 1782,	21.48	103	3.970		3.85
" since "	20.93	104	3.906		3.75
Escudo, to 1788,	20.98	52	1.957		3.76
" since "	20.42	52	1.905		3.66
Coronilla " 1800,	20.29	27	.983		3.64
SWEDEN.					
Ducat,	23.45	53	2.230		4.20

LONG OR LINEAR MEASURE.

	Standard of purity in carats.	Standard weight in grains.	Par value in Federal money.	Circulating value in Federal money.	Par value per grain. cts.
SWITZERLAND.					
BERNE, Ducat, double in pro.,	23.53	47	$1.984		4.22
BERNE, Pistole,	21.62	117½	4.558		3.88
GENEVA, Pistole,	21.87	87¾	3.443		3.92
" " (old)	21.51	103¼	3.985		3.85
ZURICH, Ducat, double in pro.,	23.50	53½	2.256		4.21
TURKEY.					
Misseir, half in pro. 1820,	15.88	36¼	1.040		2.84
Sequin fonducli,	19.25	53	1.830		3.45
Yeermeeblekblek,	22.88	73¾	3.027		4.10

NOTE.— The Milled Dollars, or *Pesos* (silver) of Spain, Mexico, Peru, Chili, and Central America, and the re-stamped of Brazil, weighing not less than 415 grains, and of 10 oz. 15 dwts. fine, are received by the United States Government at $1.00 each.
The Five Franc silver pieces of France, of 10 oz. 16 dwts. fine, and weighing 384 grains, are received at 93 cents each.
The standard silver coins of Great Britain are 11 oz. 2 dwts. fine.

LONG OR LINEAR MEASURE.—U. S.

STANDARD. — A brass rod, the length of which, at 62° Fahrenheit, is $\frac{36.9999}{39.1393}$ that of a pendulum beating seconds in *vacuo*, at the level of the sea, at the latitude of London, $= \frac{36.9999}{39.1013}$ at 32° Fah., at the gravitation at New York, $=$ the Yard.

6 points	= 1 line.	5½ yards (16½ ft.)	= 1 rod.
12 lines (72 points)	= 1 inch.	40 rods (220 yds.)	= 1 furlong.
12 inches	= 1 foot.	8 fur. (5280 feet)	= 1 stat. mile.
3 feet (36 inches)	= 1 yard.		

SPECIAL, FOR CLOTH.

2¼ inches	= 1 nail.	4 quarters (36 inches)	= 1 yard.
4 nails (9 inches)	= 1 quarter.		

SPECIAL, FOR LAND.

7$\frac{92}{100}$ inches	= 1 link.	100 links (66 feet)	= 1 chain.
25 links	= 1 rod.	80 chains (320 rods)	= 1 s. mile.

ENGINEER'S CHAIN.

10 inches = 1 link.
120 links (100 feet) = 1 chain.

SHOEMAKER'S MEASURE.

No. 1 is $4\frac{1}{4}$ inches in length, and each succeeding number is an addition of $\frac{1}{3}$ of an inch. No. 1 man's size = $8\frac{1}{3}$ inches.

MISCELLANEOUS.

Hair's breadth	= $\frac{1}{48}$ inch.	Fathom	= 6 feet.
Digit	= 10 lines.	Knot	= $47\frac{3}{4}$ feet.
Palm	= 3 inches.	Cable's length	= 120 fathoms.
Hand	= 4 "	Geometrical pace	= 4.4 feet.
Span	= 9 "		

12 particular things = 1 dozen.
12 dozen (144) = 1 gross.
12 gross (1728) = 1 great gross.
20 particular things = 1 score.
24 sheets of paper = 1 quire.
20 quires = 1 ream.

SQUARE OR SUPERFICIAL MEASURE.
(Length × breadth.)

144 square inches = 1 square foot.
  9 " feet = 1 " yard.
 $30\frac{1}{4}$ " yards = 1 " rod.
 40 " rods = 1 rood.
  4 roods = 1 acre.

SPECIAL, FOR LAND.

$62\frac{4}{5}\frac{5}{2}\frac{4}{5}$ square inches = 1 square link.
10000 " links = 1 " chain.
  10 " chains = 1 acre.

Square rod = $272\frac{1}{4}$ square feet.

Rood = { 1210 " yards.
       { 10890 " feet.

Acre (160 square rods) = { 4840 " yards.
                         { 43560 " feet.

Square mile = { 640 acres.
            { 102400 sq. rods.

220 × 198 square feet ⎫
The square of 12.649 " rods ⎪
 "   " of 69.5701 " yards ⎬ = 1 acre.
 "   " of 208.710321 " feet ⎭

CUBIC OR SOLID MEASURE.

CIRCULAR MEASURE.

Minute, or Geographical m. (60″) $= \begin{cases} 1.152 \text{ s. miles.} \\ 6086 \text{ feet.} \end{cases}$

League $= 3$ miles.

Degree $= \begin{cases} 60 \text{ geo. miles.} \\ 69.158 \text{ s. ms.} \end{cases}$

Sign($\frac{1}{12}$ zod.) $= 30$ degrees.

Great Circle $= 360$ degrees.

Equatorial circumference of the earth $= 24897$ s. m.

Equatorial diam. $= 7925$ "
Polar diam. $= 7899$ "
Mean radius $= 3955.92$ "

NOTE. — In the expressions, *square feet* and *feet square*, there is this difference; viz., the former expresses an area in which there are as many square feet as the *number* named, and the latter an area in which there are as many square feet as the *square* of the number named. The former particularizes no *form* of area, the latter asserts a *square* form.

CUBIC OR SOLID MEASURE. — U. S.
(*Length* × *breadth* × *depth*.)

Cubic foot, 1728 cu. inches $= \begin{cases} 1.273 \text{ cylindrical feet.} \\ 2200 \text{ " inches.} \\ 3300 \text{ spherical "} \\ 6600 \text{ conical "} \end{cases}$

Cylindrical foot, 1728 " inches $= \begin{cases} 0.785398 \text{ cubic feet.} \\ 1357.2 \text{ " inches.} \\ 2592 \text{ spherical "} \\ 5184 \text{ conical "} \end{cases}$

27 cubic feet $= 1$ cubic yard.
40 " of round timber $= 1$ ton.
42 " of shipping " $= 1$ ton.
50 " of hewn " $= 1$ ton.
128 " $= 1$ cord.

Cubic foot of pure water, at the maximum density at the level of the sea, (39°.83, barometer 30 inches) $= \begin{cases} 62\frac{1}{2} \text{ avoirdupois pounds.} \\ 1000 \text{ " ounces.} \end{cases}$

Cylindrical foot $= \begin{cases} 49.1 \text{ " pounds.} \\ 785.4 \text{ " ounces.} \end{cases}$

Cubic inch $= \begin{cases} 0.036169 \text{ " pounds.} \\ 0.5787 \text{ " ounces.} \\ 253.1829 \text{ grains.} \end{cases}$

Cylindrical inch $= \begin{cases} 0.028415 \text{ avd. pounds.} \\ 0.4546 \text{ " ounces.} \end{cases}$

Pound $= 27.648$ cubic inches.
" distilled $= 27.7015$ " "
Cubic inch " $= 252.6839$ grains.
Pound at 62°, distilled $= 27.7274$ cub. inches.
Cubic inch at 62°, " $= 252.458$ grains.
" " 39°.83, in *vacuo* $= 253.0864$ "

Cubic foot of salt water (sea) weighs 64.3 pounds.

GENERAL MEASURE OF WEIGHT.—U. S.

AVOIRDUPOIS.

STANDARD. — The pound is the weight, taken in air, of 27.7015 cubic inches of distilled water at its maximum density, (39°.83 F., the barometer being at 30 inches) = 27.7274 cubic inches of distilled water at 62° = 7000 Troy grains.

27$\frac{11}{32}$ grains = 1 dram.
16 drams (437$\frac{1}{2}$ grs.) = 1 ounce.
16 ounces (7000 grs.) = 1 pound.

SPECIAL — GROSS.

28 pounds = 1 quarter.
4 quarters } = { 1 quintal.
112 pounds } = { 1 cwt.
20 cwt. = 1 ton.

SPECIAL — DIAMOND.

16 parts = 1 grain = 0.8 troy gr.
4 grs. = 1 carat = 3.2 " "

SPECIAL — TROY.

(Exclusively for gold and silver bullion, precious stones, and gold, silver and copper coins, and with reference to their monetary value only.)

24 grains = 1 pennyw't.
20 dwts. (480 grs.) = 1 ounce.
12 oz. (5760 grs.) = 1 pound.

SPECIAL — APOTHECARIES'.

(Exclusively for compounding medicines, for recipes and prescriptions.)

20 grains = 1 scruple, ∋.
3 scruples = 1 dram, ℨ.
8 drams (480 g.) = 1 ounce, ℥.
12 oz. (5760 g.) = 1 pound, ℔.

1 lb. avoir. = 1$\frac{21}{144}$ lbs. troy.
1 lb. troy = $\frac{144}{175}$ lbs. avoir.
1 oz. avoir. = $\frac{175}{192}$ oz. troy.
1 oz. troy = 1$\frac{17}{175}$ oz. avoir.

NOTE. — The comparative value of diamonds of the same quality is as the square of their respective weights. A diamond of fair quality, weighing 1 carat in the rough state, is estimated worth about 9\frac{50}{100}$; and it will require one of twice that weight to make one when worked down equal to 1 carat in weight. Hence, to determine the value of a wrought diamond of any given number of carats: — *Rule.* — Double the weight in carats and multiply the square by 9.50. Thus, the value of a wrought diamond, weighing 2 carats, is 2 + 2 = 4 × 4 = 16 × 9.50 = $152.

LIQUID MEASURE.—U. S.

The "Wine" or "Winchester" Gallon, of 231 cubic inches capacity, is the Government or Customs gallon of the United States for all liquids, and the legal gallon of each state in which no law exists fixing a state or statute gallon of its own. It contains 58372$\frac{1}{2}$ grains of distilled water at 39°.83, the barometer being at 30 inches.

4 gills = 1 pint, 2 pints = 1 quart.
4 quarts, or 231 cubic in. } = { 1 gallon.
0.13368 cub. ft., 294.1176 cyl. in. } = { 8.355 av'd. lbs. pure water.

DRY MEASURE.   29

Liquid gallon of the ⎫   ⎧ 0.128 cubic foot.
State of New York,* ⎬ = ⎨ 221.184 "   in.
281.62 cylindric in. ⎭   ⎩ 8 avoid. lbs. pure water
                             at 39°.83, b. 30 in.

Barrel   = 31½ gallons. | Puncheon   = 84 gallons.
Tierce   = 42   " | Pipe or Butt = 126 "
Hogshead = 63   " | Tun   = 252 "

Imperial gallon, ⎬ = ⎨ 10 av'd lbs. distilled water
277.274 cub. in. ⎭   ⎩   at 62° F., b. 30 in.

Ale gallon, ⎬ = ⎨ 10⅕ av'd lbs. pure water
282 cub. in. ⎭   ⎩   at 39°.83, b. 30 in.

                 ⎧ 0.8331 Imperial gallon.
1 Wine gallon = ⎨ 0.8191 Ale   "
                 ⎩ 0.10742 W. bushel.

1 Imperial gallon   =   1.2   Wine gallons.

DRY MEASURE.—U. S.

The "Winchester Bushel," so called, of $2150\frac{42}{100}$ cubic inches capacity, is the Government bushel of the United States, and the legal bushel of each state having no special or statute bushel of its own. The standard Winchester bushel measure is a cylindrical vessel having an outside diameter of 19½ inches, an inside diameter of 18½ inches, and an inside depth of 8 inches. The standard "heaped" or "coal" bushel of England was this measure heaped to a true cone 6 inches high, the base being 19½ inches, or equal to the outside diameter of the measure. Its ratio to the even bushel was, therefore, as 1.28, nearly, to 1. The present "Imperial" measure of England has the same outside diameter and the same depth as the Winchester, and an internal diameter of 18.8 inches, and the same height of cone is retained for forming the heaped bushel. Its ratio, therefore, to the even bushel is a trifle less than was that of the Winchester. In the United States the "heaped bushel" is usually estimated at 5 even pecks, or as 1.25 to 1 of the standard even bushel, which, if taken as

---

* By enactment of the Legislature of the State of New York, this gallon ceased to be the legal gallon of that State, April 11, 1852; and the United States Government gallon, of 231 cubic inches capacity, was adopted in its stead.

the rule, requires a cone on the Winchester measure of 5.4 inches to equal the heaped Winchester bushel.

4 gills	=	1 pint.
2 pints	=	1 quart.
4 quarts	=	{ 1 gallon or half peck.
8 quarts	=	1 peck.
4 pecks ⎫ 2150.42 cubic in. ⎪ 1.244456 " ft. ⎬ 1.5844 cyl. " ⎭	=	⎧ 1 bushel. ⎪ 2738 cyl. in. ⎨ 77.7785 av'd lbs. ⎪ pure water.
Bushel of the ⎫ State of New York,* ⎬ 2816.1955 cyl. in. ⎭	=	⎧ 1.28 cubic feet. ⎨ 2211.84 " in. ⎩ 80 av'd lbs. pure water.
Bushel of Connecticut,†	=	⎧ 1.272 cubic feet. ⎨ 2198 " in. ⎩ 79.50 av'd lbs. pure water.
Heaped Win. bushel ⎫ 1.28—even " " ⎬	=	{ 2747.7 cubic in. { 1.59 cubic ft.
Imperial bushel	=	2218.192 " in.
Chaldron	=	36 Winch. heaped bushels.
1 Winchester bushel	=	{ 0.9694 Imperial bushel. { 9.3092 Wine gallons.
1 Imperial bushel	=	1.0315 Winchester bushels.

NOTE.—The Imperial bushel, mentioned above, is the present legal bushel of Great Britain; and the Imperial gallon, mentioned on the preceding page, is the present legal gallon of Great Britain, for all liquids. The gallon for liquids is the same as the gallon for dry measure. Eight Imperial gallons make one bushel. The subdivisions of the gallon and the bushel, and their denominations, are the same as in the Winchester measures. In Great Britain, in addition to the denominations of dry measure used in the United States, the

Strike,	= 2 bushels.	Last,	= 80 bushels.
Coomb,	= 4 "	Sack of corn,	= 3 "
Quarter,	= 8 "	Bole of corn,	= 6 "
Wey or load,	= 40 "	Last of gunpowder,	= 42 barrels.

* This bushel ceased to be the legal bushel of this State April 11, 1852, and the United States Government bushel, of $2150\frac{42}{100}$ cubic inches capacity, was adopted as the legal bushel in its stead.

† This bushel is now, January, 1852, no longer the legal bushel of this State, and the standard Winchester bushel is adopted in its stead.

SECTION II.

MISCELLANEOUS FACTS, CALCULATIONS, AND PRACTICAL MATHEMATICAL DATA.

SPECIFIC GRAVITIES.

The specific gravity of a body is its weight relative to the weight of an equal bulk of pure water at the maximum density, (39°.83, b. 30 in.) the water being taken as 1., a cubic foot of which weighs 1000 avoirdupois ounces, or 62½ lbs. The specific gravity, therefore, of any body multiplied by 1000, or, which is the same thing, the decimal being carried to three places of figures, or thousands, as in the following TABLES, the whole taken as an integer equals the number of ounces in a cubic foot of the material: multiplied by 62.5, or considered an integer and divided by 16, it equals the number of pounds in a cubic foot; and multiplied by .036169, or taken as an integer and divided by 27648, it equals the decimal fraction of a pound per cubic inch; by which, it is readily seen, the specific gravity of a commodity being known, its weight per any given bulk is easily and accurately ascertained; as, also, its specific gravity, the weight and bulk being known. The weight of any one article relative to that of any other, is as its respective specific gravity to the specific gravity of the other.

METALS.	Specific gravity.		Specific gravity.
Antimony,	6.712	Gold, pure, hammered,	19.546
Arsenic,	5.810	Iridium,	15.363
Bismuth,	9.823	Iron, cast,	7.209
Bronze,	8.700	" wrought,	7.787
Brass, best,	8.504	Lead,	11.352
Copper, cast,	8.788	Mercury, 32°,	13.598
" wire-drawn,	8.878	" 60°,	13.580
Cadmium,	8.604	" —39°,	15.000
Cobalt,	7.700	Manganese,	8.013
Chromium,	5.900	Molybdenum,	8.611
Glucinium,	3.000	Nickel,	8.280
Gold, pure, cast,	19.258	Osmium,	10.000

SPECIFIC GRAVITIES.

	Specific gravity.		Specific gravity.
Platinum, cast,	19.500	Granite, red,	2.625
" hammered,	20.337	" Lockport,	2.655
" rolled,	22.069	" Quincy,	2.652
Potassium, 60°,	0.865	" Susquehanna,	2.704
Palladium,	11.870	Grindstone,	2.143
Rhodium,	11.000	Gypsum, opaque,	2.168
Silver, pure, cast,	10.474	Hone, white,	2.876
" hammered,	10.511	Hornblende,	3.600
Sodium,	0.970	Ivory,	1.822
Steel, soft,	7.836	Jasper,	2.690
" tempered,	7.818	Limestone, green,	3.180
Tin, cast,	7.291	" white,	3.156
Tellurium,	6.115	Lime, compact,	2.720
Tungsten,	17.600	" foliated,	2.837
Titanium,	4.200	" quick,	0.804
Uranium,	9.000	Loadstone,	4.930
Zinc, cast,	6.861	Magnesia, hyd.,	2.333
		Marble, common,	2.686
STONES AND EARTHS.		" white Ital.	2.708
Alabaster, white,	2.730	" Rutland, Vt.,	2.708
" yellow,	2.699	" Parian,	2.838
Amber,	1.078	Nitre, crude,	1.900
Asbestos, starry,	3.073	Pearl, oriental,	2.650
Borax,	1.714	Peat, hard,	1.329
Bone, ox,	1.656	Porcelain, China,	2.385
Brick,	1.900	Porphyra, red,	2.766
Chalk, white,	2.782	" green,	2.675
Charcoal,	.441	Quartz,	2.647
" triturated,	1.380	Rock Crystal,	2.654
Cinnabar,	7.786	Ruby,	4.283
Clay,	1.934	Stone, common,	2.520
Coal, bitum. avg.,	1.270	" paving,	2.416
" anth. "	1.520	" pumice,	0.915
Coral, red,	2.700	" rotten,	1.981
Earth, loose,	1.500	Salt, common, solid,	2.130
Emery,	4.000	Saltpetre, refined,	2.090
Feldspar,	2.500	Sand, dry,	1.800
Flint, white,	2.594	Serpentine,	2.430
" black,	2.582	Shale,	2.600
Garnet,	4.085	Slate,	2.672
Glass, flint,	2.933	Spar, fluor,	3.156
" white,	2.892	Stalactite,	2.324
" plate,	2.710	Talc, black,	2.900
" green,	2.642	Topaz,	4.011

SPECIFIC GRAVITIES. 33

	Specific gravity.
SIMPLE SUBSTANCES, *neither metallic nor gaseous.*	
Boron,	1.968
Bromine,	2.970
Carbon,	3.521
Iodine,	4.943
Phosphorus,	1.770
Selenium,	4.320
Silicon,	1.184
Sulphur,	1.990
WOODS, *(dry.)*	
Apple,	0.793
Alder,	.800
Ash,	.760
Beech,	.696
Birch,	.720
Box, French,	1.328
" Dutch,	.912
Cedar,	.561
Cherry,	.715
Chestnut,	.610
Cocoa,	1.040
Cork,	.240
Cypress,	.644
Ebony, American,	1.331
" foreign,	1.290
Elm,	.671
Fir, yellow,	.657
" white,	.569
Hacmetac,	.592
Hickory, red,	.900
Lignum vitæ,	1.333
Larch,	.544
Logwood,	.913
Mahogany, Spanish, best,	1.065
" " com.,	.800
" St. Domingo,	.720
Maple, red,	.750
Mulberry,	.897
Oak, live,	1.120
" white,	.785
Orange,	.705
Pear,	.661
Pine, white,	.554

	Specific gravity.
Pine, yellow,	.568
Poplar, white,	.383
Plum,	.785
Quince,	.705
Spruce, white,	.551
Sassafras,	.482
Sycamore,	.604
Walnut,	.671
Willow,	.585
Yew, Spanish,	.807
" Dutch,	.788
Highly seasoned Am.	
Ash, white,	.722
Beech,	.624
Birch,	.526
Cedar,	.452
Cherry,	.606
Cypress,	.441
Elm,	.600
Fir,	.491
Hickory, red,	.838
Maple, hard,	.560
Oak, white, upland,	.687
" James River,	.759
Pine, yellow,	.541
" pitch,	.536
" white,	.473
Poplar, (tulip,)	.587
Spruce, white,	.465
GUMS, FATS, &c.	
Asphaltum,	{ .905 / 1.650 }
Beeswax,	.965
Butter,	.942
Camphor,	.988
Gamboge,	1.222
Gunpowder,	.900
" shaken,	1.000
" solid,	{ 1.550 / 1.800 }
Gum, Arabic,	1.454
" Caoutchouc,	.933
" Mastic,	1.074

SPECIFIC GRAVITIES.

	Specific gravity.		Specific gravity.
Honey,	1.450	Wine, champagne,	.997
Ice,	.930	" claret,	.994
Indigo,	1.009	" port,	.997
Lard,	.941	" sherry,	.992
Pitch,	1.150		
Rosin,	1.100		
Spermaceti,	.943	**ELASTIC FLUIDS:**	
Starch,	1.530		
Sugar, dry,	1.606	The measure of which is atmospheric air, at 60°, b. 30 in., its assumed gravity 1; one cubic foot of which weighs 527.04 grains, = .305 of a grain per cubic inch. It is, at this temperature and density, to pure water at the maximum density, as .0012046 to 1, or as 1 to 830.1.	
Tallow,	.938		
Tar,	1.015		

LIQUIDS.

Acid, acetic,	1.062	**SIMPLE OR ELEMENTARY GASES.**	
" citric,	1.034	Hydrogen,	.0689
" fluoric,	1.060	Oxygen,	1.1025
" nitric,	1.485	Nitrogen,	.9760
" nitrous,	1.420	Fluorine,	
" sulphuric,	1.846	Chlorine,	2.470
" muriatic,	1.200	Carbon, vapor of,	} .422
" silicic,	2.660	(*theoretically,*)	
Alcohol, anhyd.	.794		
" 90 %	.834		
Beer,	1.034	**COMPOUND GASES.**	
Blood, human,	1.054	Ammoniacal,	.591
Camphene, pure,	.863	Carbonic acid,	1.525
Cider, whole,	1.018	" oxide,	.763
Ether, sulph.,	.715	Carbureted hydrogen,	.559
" nitric,	.908	Chloro-carbonic,	3.389
Milk, cow's,	1.032	Cyanogen,	1.818
Molasses, 75 %	1.400	Muriatic acid gas,	1.247
Oils, linseed,	.934	Nitrous acid gas,	3.176
" olive,	.917	Nitrous oxide gas,	1.040
" rapeseed,	.927	Olefiant,	.982
" sassafras,	1.090	Phosphureted hydrogen,	1.185
" turpentine, com.,	.875	Sulphureted "	1.177
" sperm, pure,	.874	Steam, 212°	.484
" whale, p'f'd,	.923	Smoke, of wood,	.900
Proof spirits,	.925	" of coal,	.102
Vinegar,	1.025	Vapor, of water,	.623
Water, pure,	1.000	" of alcohol,	1.613
" sea,	1.026	" of spirits turpentine,	5.013
" Dead sea,	1.240		

WEIGHT PER BUSHEL — BARREL — GALLON, &C.

Weight per Bushel (even Winchester) of different Grains, Seeds, &c.

Articles.	lbs.	Articles.	lbs.
Barley, (N. E. 47 lbs.)	48	Hemp seed,	40
Beans,	64	Oats,	32
Buckwheat,	46	Peas,	64
Blue-grass seed	14	Rye,	56
Corn,	56	Salt, T. I.,	80
Cranberries,		" boiled,	56
Clover seed,	60	Timothy seed,	46
Dried Apples,	22	Wheat,	60
" Peaches,	33	Potatoes, h'p'd,	60
Flax seed, (N. E. 52 lbs.)	56		

Weight per Barrel (Legal or by Usage) of different Articles.

Flour,	196 lbs.	Cider, in Mass.,	32 gals.
Boiled Salt,	280 "	Soap,	256 lbs.
Beef,	200 "	Raisins,	112 "
Pork,	200 "	Anchovies,	30 "
Pickled Fish,	200 "	Lime,	
" " in } Massachusetts, }	30 gls.	Ground Plaster, Hydraulic Cement,	300 "

A Gallon of Train Oil weighs	7¾ lbs.
A " " Molasses, standard, (75 per cent.,)	11⅔ "
A Puncheon of Prunes,	1120 "
A Firkin of Butter, (legal,)	56 "
A Keg of powder,	25 "
A Hogshead of Salt is	8 bush.
A Perch of Stone = 24¾ cubic feet.	
A Gallon of Alcohol, 90 per cent., weighs	6.965 lbs.
A " " Proof Spirits, "	7.732 "
A " " Wine, (average,) "	8.3 "
A " " Sperm Oil, "	7.33 "
A " " Whale " p'f'd, "	7.71 "
A " " Olive " "	7.66 "
A " " Spirits Turpentine, "	7.31 "
A " " Camphene, pure, "	7.21 "

Weight of Coals, &c., broken to the medium size, per Measure of Capacity.

The average weight of Bituminous Coals, broken as above, is about 62 per cent. that of a bulk of equal dimensions in the solid mass, or of the specific gravity of the article; that of Anthracite is about 57 per cent.

Average weight per cubic foot.	lbs.	Average weight per W. Coal bushel.	lbs.
Anthracite,	54	Anthracite,	86
Bituminous,	50	Bituminous,	80
Charcoal, of pine,	18.6	Charcoal, hard wood,	30
" of hard wood,	19.02	Coke, best,	32

Practical Approximate Weight in Pounds of Various Articles.

Sand, dry, per cubic foot,	95
Clay, compact, per cubic foot,	135
Granite, " " "	165
Lime, quick, " " "	50
Marble, " " "	169
Slate, " " "	167
Peat, hard, " " "	83
Seasoned Beech Wood, per cord,	5616
" Yellow Birch Wood, per cord,	4736
" Red Maple Wood, " "	5040
" " Oak Wood, " "	6200
" White Pine Wood, " "	4264
" Hickory Wood, " "	6960
" Chestnut Wood, " "	4880

Meadow Hay, well settled, per cubic foot, $8\frac{1}{3}$ lbs., or 240 cubic feet = 2000 lbs., or $268\frac{8}{10}$ cubic feet = 1 long ton

Meadow Hay, in large old stacks, per cubic foot,	$9\frac{8}{10}$
Clover Hay, in settled bulk, " " "	$7\frac{3}{4}$
Corn on Cob, in crib, " " "	22
" shelled, in bin, " " "	45
Wheat, in bin, " " "	48
Oats, in bin, " " "	$25\frac{1}{2}$
Potatoes, in bin, " " "	$38\frac{1}{2}$
Common Brick, $7\frac{3}{4} \times 3\frac{3}{4} \times 2\frac{1}{4}$ in. " M,	4500
Front " $8 \times 4\frac{1}{4} \times 2\frac{1}{2}$ in. " "	6185

ROPES AND CABLES.

The STRENGTH of cords depends somewhat upon the fineness of the strands; — damp cordage is stronger than dry, and untarred stonger than tarred; but the latter is impervious to water and less elastic.

SILK cords have three times the strength of those of flax of equal circumference, and MANILLA has about half that of hemp.

Ropes made of IRON WIRE are full three times stronger than those of hemp of equal circumference.

White ropes are found to be most durable. The best qualities of hemp are — 1. *pearl gray;* 2. *greenish;* 3. *yellow.* A brown color has less strength.

THE BREAKING WEIGHT of a good hemp rope is 6400 lbs. per square inch, but no cordage may be counted on with safety as capable of sustaining a weight or strain above half that required to break it, and the weight of the rope itself should be included in the estimate.

THE RELIABLE STRENGTH of a good hemp cable, in pounds, is usually estimated as equal to the square of its circumference in inches $\times$ by 120. That of rope $\times$ 200. Thus, a cable of 9 inches in circumference may be relied on as having a sustaining power $= 9 \times 9 \times 120 = 9720$ lbs.

THE WEIGHT, in pounds, of a cable laid rope, per linear foot $=$ the square of its circumference in inches $\times$ .036, very nearly.

The weight, in pounds, of a linear foot of manilla rope $=$ the square of its circumference in inches $\times$ .03, very nearly. Thus, a manilla rope of three inches circumference weighs per linear foot $3 \times 3 \times .03 = \frac{27}{100}$ lbs., $= 3\frac{7}{10}$ feet per lb.

A good hemp rope stretches about $\frac{1}{8}$, and its diameter is diminished about $\frac{1}{4}$ before breaking.

WEIGHT AND STRENGTH OF IRON CHAINS.

Diameter of Wire in Inches.	Weight of 1 Foot of Chain. lbs.	Breaking Weight of Chain. lbs.	Diameter of Wire in Inches.	Weight of 1 Foot of Chain. lbs.	Breaking Weight of Chain. lbs.
$\frac{3}{16}$	0.325	2240	$\frac{5}{8}$	4.217	26880
$\frac{1}{4}$	0.65	4256	$\frac{11}{16}$	4.833	32704
$\frac{5}{16}$	0.967	6720	$\frac{3}{4}$	5.75	38752
$\frac{3}{8}$	1.383	9634	$\frac{13}{16}$	6.667	45696
$\frac{7}{16}$	1.767	13216	$\frac{7}{8}$	7.5	51744
$\frac{1}{2}$	2.633	17248	$\frac{15}{16}$	9.333	58464
$\frac{9}{16}$	3.333	21728	1	10.817	65632

Comparative Weight of Metals, Weight per Measure of Solidity, &c

	Specific Gravity.	Ratio of Comparison	Pounds in a Cubic Foot.	Inch.
Iron, wrought or rolled,	7.787	1.	486.65	.28163
Cast Iron,	7.209	.9258	450.55	.26073
Steel, soft, rolled,	7.836	1.0064	489.75	.28342
Copper, pure, "	8.878	1.1401	554.83	.32110
Brass, best, "	8.604	1.1050	537.75	.3112
Bronze, gun metal,	8.700	1.1173	543.75	.31464
Lead,	11.352	1.4579	709.50	.4106

TABLE,

Exhibiting the Weight in pounds of One Foot in Length of Wrought or Rolled Iron of any size, (cross section,) from $\frac{1}{8}$ inch to 12 inches.

SQUARE BAR.

Size in Inches.	Weight in Pounds.	Size in Inches.	Weight in Pounds.	Size in Inches.	Weight in Pounds.	Size in Inches.	Weight in Pounds.
$\frac{1}{8}$	.053	$2\frac{3}{8}$	19.066	$4\frac{3}{8}$	72.305	$7\frac{3}{4}$	203.024
$\frac{1}{4}$	.211	$2\frac{1}{2}$	21.120	$4\frac{1}{2}$	76.264	8	216.336
$\frac{3}{8}$	.475	$2\frac{5}{8}$	23.292	$4\frac{7}{8}$	80.333	$8\frac{1}{4}$	230.068
$\frac{1}{2}$	.845	$2\frac{3}{4}$	25.560	5	84.480	$8\frac{1}{2}$	244.220
$\frac{5}{8}$	1.320	$2\frac{7}{8}$	27.939	$5\frac{1}{8}$	88.784	$8\frac{3}{4}$	258.600
$\frac{3}{4}$	1.901	3	30.416	$5\frac{1}{4}$	93.168	9	273.792
$\frac{7}{8}$	2.588	$3\frac{1}{8}$	33.010	$5\frac{3}{8}$	97.657	$9\frac{1}{4}$	289.220
1	3.380	$3\frac{1}{4}$	35.704	$5\frac{1}{2}$	102.240	$9\frac{1}{2}$	305.056
$1\frac{1}{8}$	4.278	$3\frac{3}{8}$	38.503	$5\frac{5}{8}$	106.953	$9\frac{3}{4}$	321.332
$1\frac{1}{4}$	5.280	$3\frac{1}{2}$	41.408	$5\frac{3}{4}$	111.756	10	337.920
$1\frac{3}{8}$	6.390	$3\frac{5}{8}$	44.418	$5\frac{7}{8}$	116.671	$10\frac{1}{4}$	355.136
$1\frac{1}{2}$	7.604	$3\frac{3}{4}$	47.534	6	121.664	$10\frac{1}{2}$	372.672
$1\frac{5}{8}$	8.920	$3\frac{7}{8}$	50.756	$6\frac{1}{4}$	132.040	$10\frac{3}{4}$	390.628
$1\frac{3}{4}$	10.352	4	54.084	$6\frac{1}{2}$	142.816	11	408.960
$1\frac{7}{8}$	11.883	$4\frac{1}{8}$	57.517	$6\frac{3}{4}$	154.012	$11\frac{1}{4}$	427.812
2	13.520	$4\frac{1}{4}$	61.055	7	165.632	$11\frac{1}{2}$	447.024
$2\frac{1}{8}$	15.263	$4\frac{3}{8}$	64.700	$7\frac{1}{4}$	177.672	$11\frac{3}{4}$	466.684
$2\frac{1}{4}$	17.112	$4\frac{1}{2}$	68.448	$7\frac{1}{2}$	190.136	12	486.656

COMPARATIVE WEIGHT OF METALS.

To determine the weight, in pounds, of one foot in length, or of any length, of a bar of any of the following metals of form prescribed, of any size, multiply the weight in pounds, of an equal length of square rolled iron of the same size, (see table of square rolled iron,) if the weight be sought of

Iron,	Round rolled, by	. . . . .	.7854
Steel,	Square " "	. . . . .	1.0064
"	Round " "	. . . . .	.7904
Cast Iron,	Square bar, "	. . . . .	.9258
" "	Round " "	. . . . .	.7271
Copper,	Square rolled, "	. . . . .	1.1401
"	Round " "	. . . . .	.8954
Brass,	Square " "	. . . . .	1.105
"	Round " "	. . . . .	.8679
Bronze,	Square bar, "	. . . . .	1.1173
"	Round " "	. . . . .	.8775
Lead,	Square " "	. . . . .	1.4579
"	Round " "	. . . . .	1.145

The weight of a bar of any metal, or other substance, of any given length, of a *flat form*, (and any other form may be included in the rule,) is readily obtained by multiplying its cubic contents (feet or inches) by the weight (pounds, ounces, or grains) of a cubic foot or inch of the article sought to be weighed; that is—

Length × *breadth* × *thickness* × *weight per unit of measure.*

For the weight in pounds of a cubic foot or inch of different metals, see " TABLE of weights of metals per measure of solidity, &c."

OR, FOR FLAT OR SQUARE BARS,

Multiply the sectional area in inches by the length in feet, and that product, if the metal be

Wrought Iron, by	. . . . . .	3.3795
Cast " "	. . . . . .	3.1287
Steel, "	. . . . . .	3.4

EXAMPLE.—Required the weight of a bar of steel, whose length is 7 feet, breadth 2½ inches, and thickness ¾ of an inch.
    2.5 × .75 × 7 × 3.4 = 44.625 lbs. *Ans.*

EXAMPLE.—Required the weight of a cast iron beam, whose length is 14 feet, breadth 9 inches, and thickness 1½ inch.
    14 × 9 × 1.5 × 3.1287 = 591.32 lbs. *Ans.*

TABLE,

Exhibiting the weight in pounds of One Foot in Length of Round Rolled Iron of any diameter, from 1/8 inch to 12 inches.

Diameter in inches.	Weight in lbs.	Diam. in inches.	Weight in lbs.	Diam. in inches.	Weight in lbs.	Diam. in inches.	Weight in lbs.
1/8	.041	2 3/8	14.975	4 5/8	56.788	7 3/4	159.456
1/4	.165	2 1/2	16.688	4 3/4	59.900	8	169.856
3/8	.373	2 5/8	18.293	4 7/8	63.094	8 1/4	180.696
1/2	.663	2 3/4	20.076	5	66.752	8 1/2	191.808
5/8	1.043	2 7/8	21.944	5 1/8	69.731	8 3/4	203.260
3/4	1.493	3	23.888	5 1/4	73.172	9	215.040
7/8	2.032	3 1/8	25.926	5 3/8	76.700	9 1/4	227.152
1	2.654	3 1/4	28.040	5 1/2	80.304	9 1/2	239.600
1 1/8	3.360	3 3/8	30.240	5 5/8	84.001	9 3/4	252.376
1 1/4	4.172	3 1/2	32.512	5 3/4	87.776	10	266.288
1 3/8	5.019	3 5/8	34.886	5 7/8	91.634	10 1/4	278.924
1 1/2	5.972	3 3/4	37.332	6	95.552	10 1/2	292.688
1 5/8	7.010	3 7/8	39.864	6 1/4	103.704	10 3/4	306.800
1 3/4	8.128	4	42.464	6 1/2	112.160	11	321.216
1 7/8	9.333	4 1/8	45.174	6 3/4	120.960	11 1/4	336.004
2	10.616	4 1/4	47.952	7	130.048	11 1/2	351.104
2 1/8	11.988	4 3/8	50.815	7 1/4	139.544	11 3/4	366.536
2 1/4	13.440	4 1/2	53.760	7 1/2	149.328	12	382.208

To find the weight of an equilateral three-sided cast iron prism.

width of side in inches^2 × 1.354 × length in feet = weight in lbs.

EXAMPLE. — A three-sided cast iron prism is 14 feet in length, and the width of each side is 6 inches; required the weight of the prism.

$6^2 \times 1.354 \times 14 = 682.4$ lbs. *Ans.*

To find the weight of an equilateral rectangular cast iron prism.

width of side in inches^2 × 3.128 × length in feet = weight in lbs.

To find the weight of an equilateral five-sided cast iron prism.

width of side in inches^2 × 5.381 × length in feet = weight in lbs.

To find the weight of an equilateral six-sided cast iron prism.

width of side in inches^2 × 8.128 × length in feet = weight in lbs.

To find the weight of an equilateral eight-sided cast iron prism.

width of side in inches^2 × 15.1 × length in feet = weight in lbs.

To find the weight of a cast iron cylinder.

diameter in inches^2 × 2.457 × length in feet = weight in lbs.

In a quantity of cast iron weighing 125 lbs., how many cubic inches?

By tabular weight per cubic inch —

$125 \div .26073 = 479.4$ cubic inches. *Ans.*

Or, by tabular weight per cubic foot —
$$450.55 : 1728 :: 125 : 479.4 \text{ cubic inches.} \quad Ans.$$

How many cubic inches of copper will weigh as much as 479.4 cubic inches of cast iron?

By tabular weight per cubic inch —
$$.3211 : .26073 :: 479.4 : 389.27 \text{ cubic inches.} \quad Ans.$$

Or, by specific gravities —
$$8.878 : 7.209 :: 479.4 : 389.27 \text{ cubic inches.} \quad Ans.$$

Or, by tabular ratio of weight —
$$479.4 \times \frac{.9258}{1.1401} = 389.28.$$

A cast iron rectangular weight is to be constructed having a breadth of 4 inches and a thickness of 2 inches, and its weight is to be 18 lbs.; what must be its length?

$$\frac{18}{4 \times 2 \times .26073} = 8.63 \text{ inches.} \quad Ans.$$

A cast iron cylinder is to be 2 inches in diameter, and is to weigh 6 lbs.; what must be its length?

$.26073 \times .7854 = .2047$ lb. = weight of 1 cyl. inch, then

$$\frac{6}{2^2 \times .2047} = 7.327 \text{ inches.} \quad Ans.$$

A cast iron cylinder is to weigh 6 lbs., and its length is to be 7.327 inches; what must be its diameter?

$$\sqrt{\left(\frac{6}{7.327 \times .2047}\right)} = 2 \text{ inches.} \quad Ans.$$

A cast iron weight, in the form of a *prismoid*, or the *frustrum of a pyramid*, or the *frustrum of a cone*, is to be constructed that will weigh 14 lbs., and the area of one of the bases is to be 16 inches, and that of the other 4 inches; what must be the length of the weight?

$\sqrt{16 \times 4} = 8$ and $8 + 16 + 4 \div 3 = 9.33$, and $\frac{14}{9.33 \times .26073}$ = 5.75 inches. *Ans.*

Note. — For Rules in detail pertaining to the foregoing, see Geometry, *Mensuration of superficies — of solids.*

A model for a piece of casting, made of dry white pine, weighs 7 lbs.; what will the casting weigh, if made of common brass?

By specific gravities —
$$.554 : 8.604 :: 7 : 108.71 \text{ lbs.} \quad Ans.$$

Note. — As the specific gravity of the substance of which the model is composed must generally remain to some extent uncertain, calculations of this kind can only be relied on as approximate.

TABLE

Exhibiting the Weight of One Foot in Length of Flat, Rolled Iron; Breadth and Thickness in Inches, Weight in Pounds.

Br. and Th. inch.	Wei't. lbs.	Br. and Th. inch.	Wei't. lbs.	Br. and Th. inch.	Wei't. lbs.	Br. and Th. inch.	Wei't. lbs.
½ by ⅛	.211	1¼ by ⅞	3.696	1¾ by ½	2.957	2⅛ by ½	3.591
¼	.422	1	4.224	⅝	3.696	⅝	4.488
⅜	.634	1⅛	4.752	¾	4.435	¾	5.386
⅝ by ⅛	.264	1⅜ by ⅛	.581	⅞	5.175	⅞	6.284
¼	.528	¼	1.161	1	5.914	1	7.181
⅜	.792	⅜	1.742	1⅛	6.653	1⅛	8.079
½	1.056	½	2.323	1¼	7.393	1¼	8.977
¾ by ⅛	.316	⅝	2.904	1⅜	8.132	1⅜	9.874
¼	.633	¾	3.485	1½	8.871	1½	10.772
⅜	.950	⅞	4.066	1⅝	9.610	2¼ by ⅛	.950
½	1.267	1	4.647	1⅞ by ⅛	.792	¼	1.901
⅝	1.584	1⅛	5.228	¼	1.584	⅜	2.851
⅞ by ⅛	.369	1¼	5.808	⅜	2.376	½	3.802
¼	.789	1½ by ⅛	.634	½	3.168	⅝	4.752
⅜	1.108	¼	1.267	⅝	3.960	¾	5.703
½	1.478	⅜	1.901	¾	4.752	⅞	6.653
⅝	1.848	½	2.534	⅞	5.544	1	7.604
¾	2.218	⅝	3.168	1	6.336	1⅛	8.554
1 by ⅛	.422	¾	3.802	1⅛	7.129	1¼	9.505
¼	.845	⅞	4.435	1¼	7.921	1⅜	10.455
⅜	1.267	1	5.069	1⅜	8.713	1½	11.406
½	1.690	1⅛	5.703	1½	9.505	1⅝	12.356
⅝	2.112	1¼	6.337	1⅝	10.297	1¾	13.307
¾	2.534	1⅜	6.970	1¾	11.089	2⅜ by ⅛	1.003
⅞	2.957	1⅝ by ⅛	.686	2 by ⅛	.845	¼	2.006
1⅛ by ⅛	.475	¼	1.378	¼	1.690	⅜	3.010
¼	.950	⅜	2.059	⅜	2.534	½	4.013
⅜	1.425	½	2.746	½	3.379	⅝	5.016
½	1.901	⅝	3.432	⅝	4.224	¾	6.019
⅝	2.376	¾	4.119	¾	5.069	⅞	7.023
¾	2.851	⅞	4.805	⅞	5.914	1	8.026
⅞	3.326	1	5.492	1	6.759	1⅛	9.029
1	3.802	1⅛	6.178	1⅛	7.604	1¼	10.032
1¼ by ⅛	.528	1¼	6.864	1¼	8.449	1⅜	11.036
¼	1.056	1⅜	7.551	1⅜	9.294	1½	12.039
⅜	1.584	1½	8.237	1½	10.188	1⅝	13.042
½	2.112	1¾ by ⅛	.739	2⅛ by ⅛	.898	1¾	14.046
⅝	2.640	¼	1.478	¼	1.795	2	16.052
¾	3.168	⅜	2.218	⅜	2.698	2½ by ⅛	1.056

TABLE. — *Continued.*

Br. and Th. *inch.*	Weight. *lbs.*	Br. and Th. *inch.*	Weight. *lbs.*	Br. and Th. *inch.*	Weight. *lbs.*	Br. and Th. *inch.*	Weight. *lbs.*
2½ by ¼	2.112	2¾ by 1¾	16.264	3¼ by ⅝	6.865	3¾ by 1⅝	20.594
⅜	3.168	1⅞	17.426	¾	8.238	1¾	22.178
½	4.224	2	18.587	⅞	9.610	1⅞	23.762
⅝	5.280	2⅛	19.749	1	10.983	2	25.347
¾	6.336	2¼	20.911	1⅛	12.356	2¼	28.515
⅞	7.393	2⅞ by ⅛	1.214	1¼	13.729	2½	31.683
1	8.449	¼	2.429	1⅜	15.102	2¾	34.851
1⅛	9.505	⅜	3.644	1½	16.475	4 by ⅛	1.690
1¼	10.561	½	4.858	1⅝	17.848	¼	3.379
1⅜	11.617	⅝	6.073	1¾	19.221	½	6.759
1½	12.673	¾	7.287	1⅞	20.594	¾	10.139
1⅝	13.729	⅞	8.502	2	21.967	1	13.518
1¾	14.785	1	9.716	2¼	24.713	1¼	16.898
1⅞	15.841	1⅛	10.931	2½	27.459	1½	20.277
2	16.898	1¼	12.145	3½ by ⅛	1.478	1¾	23.657
2⅝ by ⅛	1.109	1⅜	13.360	¼	2.957	2	27.036
¼	2.218	1½	14.574	⅜	4.436	2¼	30.416
⅜	3.327	1⅝	15.789	½	5.914	2½	33.795
½	4.436	1¾	17.003	⅝	7.393	2¾	37.175
⅝	5.545	1⅞	18.218	¾	8.871	3	40.555
¾	6.653	2	19.432	⅞	10.350	3¼	43.934
⅞	7.762	2⅛	20.647	1	11.828	4¼ by ⅛	1.795
1	8.871	2¼	21.861	1⅛	13.307	¼	3.591
1⅛	9.980	3 by ⅛	1.267	1¼	14.785	½	7.181
1¼	11.089	¼	2.535	1⅜	16.264	¾	10.772
1⅜	12.198	⅜	3.802	1½	17.743	1	14.363
1½	13.307	½	5.069	1⅝	19.221	1¼	17.954
1⅝	14.416	⅝	6.337	1¾	20.700	1½	21.544
1¾	15.525	¾	7.604	1⅞	22.178	1¾	25.135
1⅞	16.634	⅞	8.871	2	23.657	2	28.726
2	17.742	1	10.139	2¼	26.614	2¼	32.317
2⅛	18.851	1⅛	11.406	2½	29.571	2½	35.908
2¾ by ⅛	1.162	1¼	12.673	2¾	32.528	2¾	39.498
¼	2.323	1⅜	13.941	3¾ by ⅛	1.584	3	43.089
⅜	3.485	1½	15.208	¼	3.168	3¼	46.680
½	4.647	1⅝	16.475	⅜	4.752	3½	50.271
⅝	5.808	1¾	17.743	½	6.337	4½ by ¼	3.802
¾	6.970	1⅞	19.010	⅝	7.921	½	7.604
⅞	8.132	2¼	20.277	¾	9.505	¾	11.406
1	9.294	2	22.812	⅞	11.089	1	15.208
1⅛	10.455	2½	25.345	1	12.673	1¼	19.010
1¼	11.617	3¼ by ⅛	1.373	1⅛	14.257	1½	22.812
1⅜	12.779	¼	2.746	1¼	15.842	1¾	26.614
1½	13.940	⅜	4.119	1⅜	17.426	2	30.416
1⅝	15.102	½	5.492	1½	19.010	2¼	34.218

TABLE. — *Continued.*

Br. and Th. *inch.*	Weight. *lbs.*	Br. and Th. *inch.*	Weight. *lbs.*	Br. and Th. *inch.*	Weight. *lbs.*	Br. and Th. *inch.*	Weight. *lbs.*
4½ by 2½	38.020	4¾ by 3	48.158	5¼ by ¾	13.307	5½ by 2	37.175
2¾	41.822	3¼	52.172	1	17.743	2½	46.469
3	45.624	3½	56.185	1¼	22.178	3	55.762
3¼	49.426	5 by ¼	4.224	1½	26.614	5¾ by ¼	4.858
3½	53.228	½	8.449	1¾	31.049	½	9.716
4¾ by ¼	4.013	¾	12.673	2	35.485	¾	14.574
½	8.026	1	16.898	2¼	39.921	1	19.432
¾	12.040	1¼	21.122	2½	44.356	1¼	24.290
1	16.053	1½	25.347	3	53.228	1½	29.146
1¼	20.066	1¾	29.571	5½ by ¼	4.647	1¾	34.007
1½	24.079	2	33.795	½	9.294	2	38.865
1¾	28.092	2¼	38.020	¾	13.941	2¼	43.723
2	32.106	2½	42.244	1	18.587	2½	48.581
2¼	36.119	3	46.469	1¼	23.234	3	58.297
2½	40.132	5¼ by ¼	4.436	1½	27.881	6 by ¼	5.069
2¾	44.145	½	8.871	1¾	32.528		

WEIGHT OF METALS IN PLATE.

The weight of a SQUARE FOOT *one* inch thick of

Malleable Iron . . .	= 40.554 lbs.
Com. plate " . .	= 37.761 "
Cast Iron . . .	= 37.546 "
Copper, wrought . .	= 46.240 "
" com. plate . .	= 45.312 "
Brass, plate, com. . .	= 42.812 "
Zinc, cast, pure . .	= 35.734 "
" sheet . . .	= 37.448 "
Lead, cast . . .	= 59.125 "

And for any other thickness, greater or less, it is the same in proportion; thus, a square foot of sheet copper $\frac{1}{16}$ of an inch thick $= 46.24 \div 16 = 2.89$ lbs. And 5 square feet at that thickness $= 2.89 \times 5 = 14.45$ lbs., &c. So, too, 5 square feet at 2½ inches thickness $= 46.24 \times 2.5 \times 5 = 578$ lbs.

THE AMERICAN WIRE GAUGE.

The American Wire Gauge was prepared by Messrs. Brown and Sharp, manufacturers of machinists' tools, Providence, R. I. It is graded upon geometrical principles, is rapidly becoming the standard gauge with manufacturers of wire and plate in the United States, and cannot fail to supersede the use of the Birmingham Gauge in this country.

TABLE

Showing the Linear Measures represented by Nos. American Wire Gauge and Birmingham Wire Gauge, or the values of the Nos. in the United-States Standard Inch.

No.	American Gauge. Inch.	Birm. Gauge. Inch.	No.	American Gauge. Inch.	Birm. Gauge Inch.	No.	American Gauge. Inch.	Birm. Gauge. Inch.	No.	American Gauge. Inch.	Birm. Gauge. Inch.
0000	.46000	.454	8	.12849	.165	19	.03589	.042	30	.01003	.012
000	.40964	.425	9	.11443	.148	20	.03196	.035	31	.00893	.010
00	.36480	.380	10	.10189	.134	21	.02846	.032	32	.00795	.009
0	.32486	.340	11	.09074	.120	22	.02535	.028	33	.00708	.008
1	.28930	.300	12	.08081	.109	23	.02257	.025	34	.00630	.007
2	.25763	.284	13	.07196	.095	24	.02010	.022	35	.00561	.005
3	.22942	.259	14	.06408	.083	25	.01790	.020	36	.00500	.004
4	.20431	.238	15	.05707	.072	26	.01594	.018	37	.00445	
5	.18194	.220	16	.05082	.065	27	.01419	.016	38	.00396	
6	.16202	.203	17	.04526	.058	28	.01264	.014	39	.00353	
7	.14428	.180	18	.04030	.049	29	.01126	.013	40	.00314	

Thus the DIAMETER or size of No. 4 *wire*, American gauge, is 0.20431 of an inch; Birmingham gauge, 0.238 of an inch: so the THICKNESS of No. 4 *plate*, American gauge, is 0.20431 of an inch; Birmingham gauge, 0.238 of an inch; and so for the other Nos. on the gauges respectively.

TABLE

Showing the Number of Linear Feet in One Pound, Avoirdupois, of Different Kinds of Wire; Sizes or Diameters corresponding to Nos. American Wire-gauge.

No.	Iron. Feet.	Copper. Feet.	Brass. Feet.	No.	Iron. Feet.	Copper. Feet.	Brass. Feet.
0000	1.7834	1.5616	1.6552	19	293.00	256.57	271.94
000	2.2488	1.9692	2.0872	20	396.41	347.12	367.92
00	2.8356	2.4830	2.6318	21	465.83	407.91	432.35
0	3.5757	3.1311	3.3187	22	587.35	514.32	545.13
1	4.5088	3.9482	4.1847	23	740.74	648.63	687.50
2	5.6854	4.9785	5.2768	24	934.03	817.89	866.90
3	7.1695	6.2780	6.6542	25	1177.7	1031.3	1093.0
4	9.0403	7.9162	8.3906	26	1485.0	1300.4	1378.3
5	11.400	9.9825	10.581	27	1872.7	1639.8	1738.1
6	14.375	12.588	13.342	28	2361.4	2067.8	2191.7
7	18.127	15.873	16.824	29	2977.9	2607.6	2763.8
8	22.857	20.015	21.214	30	3754.8	3287.9	3484.9
9	28.819	25.235	26.748	31	4734.2	4145.5	4694.0
10	36.348	31.828	33.735	32	5970.6	5221.2	5541.4
11	45.829	40.131	42.535	33	7528.1	6592.0	6987.0
12	57.790	50.604	53.636	34	9495.6	8314.9	8813.1
13	72.949	63.878	67.706	35	11972	10483	11111
14	91.861	80.439	85.258	36	15094	13217	14009
15	115.86	100.75	107.53	37	19030	16664	17662
16	146.10	127.94	135.60	38	24003	21018	22278
17	184.26	168.35	171.02	39	30266	26503	28091
18	232.34	203.45	215.64	40	38176	33342	35432

NOTE.— In this TABLE the iron and copper employed are supposed to be nearly pure. The specific gravity of the former was taken at 7.774; that of the latter, at 8.878. The specific gravity of the brass was taken at 8.376.

WIRE AND WIRE GAUGES.

To find the number of feet in a pound of wire of any material not given in the TABLE, *of any size, American gauge, its specific gravity being known.*

RULE. — Multiply the number of feet in a pound of iron wire of the same size by 7.774, and divide the product by the specific gravity of the wire whose length is sought; or ordinarily, for steel wire, multiply the number of feet in a pound of iron wire of the same size by 0.991.

To find the number of feet in a pound of wire of any given No., Birmingham gauge.

RULE. — Multiply the number of feet in a pound of the same kind of wire, same No., American gauge, by the size, American gauge, and divide the product by the size, Birmingham gauge.

EXAMPLE. — In a pound of copper wire No. 16, American gauge, there are 127.94 feet: how many feet are there of the same kind of wire, same No., Birmingham gauge?

$$(127.94 \times .05082) \div .065 = 100.03. \ Ans.$$

To find the weight of any given length of wire of any given No. or size, American gauge, or the length in any given weight, by help of the foregoing TABLE.

EXAMPLE. — Required the weight of 600 feet of No. 18 iron wire.

$$600 \div 232.34 = 2.5822 \text{ lbs.} = 2 \text{ lbs. } 9\tfrac{1}{3} \text{ oz., nearly. } Ans.$$

EXAMPLE. — Required the length in feet of 2½ lbs. of No. 31 brass wire.

$$4394 \times 2.5 = 10985. \ Ans.$$

Characteristics of Alloys of Copper and Zinc — Brass.

Parts by Weight.		Specific Gravity.	Color.	Denomination.
Copper.	Zinc.			
83	17	8.415	Yellowish Red.	Bath Metal.
80	20	8.448	" "	Dutch Brass.
74½	25½	8.397	Pale yellow.	Rolled Sheet Brass.
66	34	8.299	Full "	English Sheet Brass.
49½	50½	8.230	" "	German Sheet Brass.
33	67	8.284	Deep "	Watchmaker's Brass.

NOTE. — To alloys of copper and zinc, generally, there is added a small quantity of lead, which renders them the better adapted for turning, planing, or filing; and, for the same reason, to alloys of copper and tin, there is usually added a small quantity of zinc (see ALLOYS AND COMPOSITIONS).

TABLE

Showing the Weight of One Square Foot of Rolled Metals, thickness corresponding to Nos., American Wire-gauge.

Thickness. No.	Iron. Pounds.	Steel. Pounds.	Copper. Pounds.	Brass. Pounds.	Lead. Pounds.	Zinc. Pounds.
1	10.849	10.999	13.109	12.401	17.102	10.833
2	9.6611	9.7953	11.674	11.043	15.228	9.6466
3	8.6032	8.7227	10.396	9.8340	13.562	8.5903
4	7.6616	7.7680	9.2578	8.7576	12.078	7.6501
5	6.8228	6.9175	8.2442	7.7988	10.755	6.8126
6	6.0758	6.1601	7.3416	6.9450	9.5779	6.0667
7	5.4105	5.4856	6.5377	6.1845	8.5291	5.4024
8	4.8184	4.8853	5.8222	5.5077	7.5957	4.8112
9	4.2911	4.3507	5.1851	4.9050	6.7645	4.2847
10	3.8209	3.8740	4.6169	4.3675	6.0233	3.8151
11	3.4028	3.4501	4.1117	3.8896	5.3642	3.3977
12	3.0303	3.0720	3.6616	3.4638	4.7770	3.0257
13	2.6985	2.7360	3.2607	3.0845	4.2539	2.6934
14	2.4035	2.4365	2.9042	2.7473	3.7889	2.3999
15	2.1401	2.1698	2.5829	2.4463	3.3737	2.1369
16	1.9058	1.9322	2.3028	2.1784	3.0043	1.9029
17	1.6971	1.7207	2.0506	1.9399	2.6753	1.6945
18	1.5114	1.5324	1.8263	1.7276	2.3826	1.5091
19	1.3459	1.3646	1.6263	1.5384	2.1217	1.3439
20	1.1985	1.2152	1.4482	1.3700	1.8893	1.1967
21	1.0673	1.0821	1.2897	1.2300	1.6768	1.0657
22	.95051	.96371	1.1485	1.0865	1.4984	.94908
23	.84641	.85815	1.0227	.96749	1.3343	.84514
24	.75375	.76422	.91078	.86158	1.1882	.75262
25	.67125	.68057	.81109	.76728	1.0582	.67024
26	.59775	.60605	.72228	.68326	.94229	.59685
27	.53231	.53970	.64345	.60846	.83913	.53151
28	.47404	.48062	.57280	.54185	.74728	.47333
29	.42214	.42800	.51009	.48242	.66546	.42151
30	.37594	.38116	.45426	.42972	.59263	.37538

NOTE.— In calculating the foregoing TABLE, the specific gravities were taken as follows: viz., iron, 7.200; steel, 7.300; copper, 8.700; brass, 8.230; lead, 11.350; Zinc, 7.180.

TIN PLATES.

Brand Marks.	Size of Sheets in Inches.	No. of Sheets in Box.	Net Weight in lbs.	Brand Marks.	Size of Sheets in Inches.	No. of Sheets in Box.	Net Weight in lbs.
IC	14 × 14	200	140	SDXX	15 × 11	200	210
IC	14 × 10	225	112	SDXXX	15 × 11	200	231
IIC	14 × 10	225	119	SDXXXX	15 × 11	200	252
HX	14 × 10	225	147	TT	14 × 10	225	112
IX	14 × 10	225	140	" IC	12 × 12	225	119
IXX	14 × 10	225	161	" IX	12 × 12	225	147
IXXX	14 × 10	225	182	" IXX	12 × 12	225	168
IXXXX	14 × 10	225	203	" IXXX	12 × 12	225	189
IX	14 × 14	200	174	" IXXXX	12 × 12	225	210
IXX	14 × 14	200	200	" IC	20 × 14	112	112
DC	17 × 12½	100	105	" IX	20 × 14	112	140
DX	17 × 12½	100	126	" IXX	20 × 14	112	161
DXX	17 × 12½	100	147	" IXXX	20 × 14	112	182
DXXX	17 × 12½	100	168	" IXXXX	20 × 14	112	203
DXXXX	17 × 12½	100	189	Ternes IC	20 × 14	112	112
SDC	15 × 11	200	168	" IX	20 × 14	112	140
SDX	15 × 11	200	189				

NOTE.—The above TABLE includes all the regular sizes and qualities of tin plates, except "*wasters*." Other sizes, such as 10 × 10, 11 × 11, 13 × 13, &c., of the different brands, are often imported into the United States to order.

Common English Sheet Iron, Nos. 10 to 28, Birmingham gauge, widths from 24 to 36 inches.

R. G. Sheet Iron, Nos. 10 to 30, Birmingham gauge, widths from 24 to 36 inches.

American Puddled Sheet Iron, Nos. 22 to 28, Birmingham gauge, widths from 24 to 36 inches.

Russia Sheet Iron, Nos. 16 to 8 inclusive, Russia gauge, sheets 28 × 56 inches.

Sheet Zinc, Nos. 16 to 8, Liege gauge, widths from 24 to 40 inches; length 84 inches.

Copper Sheathing, 14 × 48 inches, 14 to 32 oz. (even numbers), per square foot.

Yellow Metal, in sheets, 48 × 14 inches, 14 to 32 oz. (even numbers), per square foot.

TABLE

Showing the Capacity, in Wine Gallons, of Cylindrical Cans, of different diameters, at One Inch depth. Diameter in Inches.

Diam'r. *inches.*	Gallons.	Diam'r. *inches.*	Gallons.	Diam'r. *inches.*	Gallons.	Diam'r. *inches.*	Gallons.
6	.1224	12¼	.5102	18½	1.164	24¾	2.083
6¼	.1328	12½	.5313	18¾	1.195	25	2.125
6½	.1437	12¾	.5527	19	1.227	25¼	2.167
6¾	.1549	13	.5746	19¼	1.260	25½	2.211
7	.1666	13¼	.5969	19½	1.293	25¾	2.254
7¼	.1787	13½	.6197	19¾	1.326	26	2.298
7½	.1913	13¾	.6428	20	1.360	26¼	2.343
7¾	.2042	14	.6664	20¼	1.394	26½	2.388
8	.2176	14¼	.6904	20½	1.429	26¾	2.433
8¼	.2314	14½	.7149	20¾	1.464	27	2.479
8½	.2457	14¾	.7397	21	1.499	27¼	2.524
8¾	.2603	15	.7650	21¼	1.535	27½	2.571
9	.2754	15¼	.7907	21½	1.572	27¾	2.518
9¼	.2909	15½	.8169	21¾	1.608	28	2.666
9½	.3069	15¾	.8434	22	1.646	28¼	2.713
9¾	.3233	16	.8704	22¼	1.683	28½	2.762
10	.3400	16¼	.8978	22½	1.721	28¾	2.810
10¼	.3572	16½	.9257	22¾	1.760	29	2.859
10½	.3749	16¾	.9539	23	1.799	29¼	2.909
10¾	.3929	17	.9826	23¼	1.837	29¾	3.009
11	.4114	17¼	1.0120	23½	1.877	30	3.060
11¼	.4303	17½	1.0410	23¾	1.918	30½	3.163
11½	.4497	17¾	1.0710	24	1.958	31	3.264
11¾	.4694	18	1.1020	24¼	1.999	31½	3.374
12	.4896	18¼	1.1320	24½	2.041	32	3.482

Applications of the foregoing TABLE.

EXAMPLE. — A cylindrical can is 11¼ inches in diameter, and its depth is 18⅗ inches; required its capacity.

.4303 × 18⅗ = 8 gallons. *Ans.*

EXAMPLE. — The diameter of a can containing oil is 26½ inches, and the oil is 14½ inches in depth. How many gallons are there of the oil?

2.388 × 14½ = 34.6 gallons. *Ans.*

EXAMPLE. — A can is to be constructed that will hold just 36 gallons, and its diameter is to be 18 inches; what must be its depth?

36 ÷ 1.102 = 32⅔ inches. *Ans.*

CAPACITY OF CYLINDRICAL CANS. 51

EXAMPLE. — A cylindrical can is to be constructed that shall have a depth of 15 inches and a capacity of just 5 gallons; what must be its diameter?

$5 \div 15 = .3333$ = capacity of can in gallons for each inch of depth; and against .3333 gallon in the table, or the quantity in gallons nearest thereto, is 10 inches, the required, or nearest tabular diameter. *Ans.*

NOTE. — The table is not intended to meet demands of the nature of the one contained in the last example, with accuracy, unless the fractional part of the diameter, if there be a fractional part, is $\frac{1}{4}$, $\frac{1}{2}$ or $\frac{3}{4}$ inch. As, however, the diameter opposite the tabular gallon nearest the one sought, even at its greatest possible remove, can be but about $\frac{1}{2}$ inch from the diameter required, we can, by inspection, determine the diameter to be taken, or true answer to the inquiry, sufficiently near for practical purposes, be the fraction what it may. Or, to throw the demand into a mathematical formula: As the tabular gallon nearest the one sought is to the diameter opposite, so is the tabular gallon required to the required diameter, nearly. Thus, in answer to the last query,

.3400 : 10 :: .3333 : 9.8 inches, the required or true diameter, nearly.

For a mathematical formula strictly applicable to this question, see GAUGING

Or, for a formula more strictly geometrical, we have

$$\sqrt{\frac{\text{Capacity} \times 231}{\text{Depth} \times .7854}} = \text{diameter.}$$

The true diameter, therefore, for the supposed can, is

$$\sqrt{\frac{231 \times 5}{15 \times .7854}} = 9.9- \text{ inches.}$$

WEIGHT OF PIPES.

The weight of ONE FOOT IN LENGTH of a pipe, of any diameter and thickness, may be ascertained by multiplying the square of its exterior diameter, in inches, by the weight of 12 *cylindrical* inches of the material of which the pipe is composed, and by multiplying the square of its interior diameter, in inches, by the same factor and subtracting the product of the latter from that of the former, — the remainder or difference will be the weight. This is evident from the fact that the process obtains the weight of two solid cylinders of equal length, (one foot,) the diameter of one being that of the pipe, and the other that of the vacancy, or bore. For very large pipes, the dimensions may be taken in feet, and the weight of a cylindrical foot of the material used as the factor, or multiplier, if desired.

The weight of 12 *cylindrical* inches (length 1 foot, diameter 1 inch) of

Malleable Iron   = 2.6543 lbs.
Cast Iron   = 2.4573 "
Copper, wrought,   = 3.0317 "
Lead   = 3.8697 "
Cast Iron — 1 *cyl.* foot — = 353.86 "

Therefore — EXAMPLE. — Required the weight of a copper pipe whose length is 5 feet, exterior diameter $3\frac{1}{4}$ inches, and interior diameter 3 inches.

$3\frac{1}{4} = \frac{13}{4} \times \frac{13}{4} = 10.5625 \times 3.0317 = 32.022 \; +$
$\phantom{3\frac{1}{4} =} 3 \times 3 = 9 \times 3.0317 = 27.285 \; +$
$\phantom{3\frac{1}{4} = 3 \times 3 = 9 \times 3.0317} Ans. \; \overline{4.737} \times 5 = 23.685 \text{ lbs.}$

EXAMPLE. — Required the weight of a cast iron pipe, whose length is 10 feet, exterior diameter 38 inches, and interior diameter 3 feet.
$38^2 \times 2.4573 - 36^2 \times 2.4573 = 363.68 \times 10 = 3636.8$ lbs. *Ans.*
Or, $\overline{38^2 - 36^2} = 148 \times 2.4573 = 363.68 \times 10 = 3636.8$ lbs. *Ans.*

EXAMPLE. — Required the weight of a lead pipe, whose length is 1200 feet, exterior diameter $\frac{7}{8}$ of an inch, and interior diameter $\frac{9}{16}$ of an inch.
$\frac{7}{8} \times \frac{7}{8} = \frac{49}{64} = .765625$, and $\frac{9}{16} \times \frac{9}{16} = \frac{81}{256} = .316406$, and
$.765625 - .316406 = .449219 \times 3.8697 \times 1200 = 2086$ lbs. *Ans.*

EXAMPLE. — The length of a cast-iron cylinder is 1 foot, its exterior diameter is 12 inches, and its interior diameter 10 inches; required its weight.
$12^2 - 10^2 = 44 \times 2.4573 = 108.12$ lbs. *Ans.*
Or, $144 : 353.86 :: 44 : 108.12$ lbs. *Ans.*

WEIGHT OF PIPES.

The following TABLE *exhibits the coefficients of weight, in pounds, of one foot in length, of various thicknesses, of different kinds of pipe, of any diameter whatever.*

Thickness in Inches.	Wrought Iron.	Copper.	Lead.
$\frac{1}{32}$	.332	.379	.484
$\frac{1}{16}$	.664	.758	.9675
$\frac{3}{32}$	.995	1.137	1.451
$\frac{1}{8}$	1.327	1.516	1.935
$\frac{5}{32}$	1.658	1.894	2.417
$\frac{3}{16}$	1.99	2.274	2.901
$\frac{7}{32}$	2.323	2.653	3.386
$\frac{1}{4}$	2.654	3.032	3.87
$\frac{5}{16}$	3.318	3.79	4.837
$\frac{3}{8}$	3.981	4.548	5.805

CAST IRON.

Thickness.	Factor.	Thickness.	Factor.	Thickness.	Factor.
$\frac{3}{16}$	1.842	$\frac{5}{8}$	6.143	$1\frac{1}{4}$	12.287
$\frac{1}{4}$	2.457	$\frac{3}{4}$	7.372	$1\frac{1}{2}$	14.744
$\frac{3}{8}$	3.686	$\frac{7}{8}$	8.0	$1\frac{3}{4}$	17.201
$\frac{1}{2}$	4.901	1	9.829	2	19.659

To obtain the weight of pipes by means of the above TABLE—

RULE. — Multiply the diameter of the pipe, taken from the interior surface of the metal on the one side to the exterior surface on the opposite, (interior diameter + thickness,) in inches, by the number in the table under the respective metal's name, and opposite the thickness corresponding to that of the pipe — the product will be the weight, in pounds, of ONE foot in length of the pipe, and that product multiplied by the length of the pipe, in feet, will give the weight for any length required.

EXAMPLE. — Required the weight of a copper pipe whose length is 5 feet, interior diameter and thickness $3\frac{1}{8}$ inches, and thickness $\frac{1}{8}$ of an inch.
$3\frac{1}{8} = \frac{25}{8} = 3.125 \times 1.516 \times 5 = 23.687$ lbs. *Ans.*

EXAMPLE. — Required the weight of a cast iron pipe, 10 feet in length, whose interior diameter is 3 feet, and whose thickness is 1 inch.
$36 + 1 = 37 \times 9.829 \times 10 = 3636.73$ lbs. *Ans.*

WEIGHT OF CAST IRON AND LEAD BALLS.

To find the weight of a sphere or globe of any material —
RULE. — Multiply the cube of the diameter, in inches, or feet, by the weight of a *spherical* inch or foot of the material.
The weight of a spherical inch of

$$\text{Cast Iron} = .1365 \text{ lbs.}$$
$$\text{Lead} = .215 \text{ "}$$

Therefore — EXAMPLE. — Required the weight of a leaden ball whose diameter is $\frac{1}{4}$ of an inch.

$$\tfrac{1}{4} \times \tfrac{1}{4} \times \tfrac{1}{4} = \tfrac{1}{64} = .015625 \times .215 = .00336 \text{ lb.} \quad Ans.$$

EXAMPLE. — Required the weight of a cast iron ball whose diameter is 8 inches.

$$8^3 \times .1365 = 69.888 \text{ lbs.} \quad Ans.$$

EXAMPLE. — How many leaden balls, having a diameter $\frac{1}{4}$ of an inch each, are there in a pound?

$$1 \div .00336 = \tfrac{100000}{336} = 298. \quad Ans.$$

EXAMPLE. — What must be the diameter of a cast iron ball, to weigh 69.888 lbs?

$$69.888 \div .1365 = \sqrt[3]{512} = 8 \text{ inches.} \quad Ans.$$

EXAMPLE. — What must be the diameter of a leaden ball to equal in weight that of a cast iron ball, whose diameter is 8 inches?

[Lead is to cast iron as .215 to .1365, as 1.575 to 1.]

$$8^3 = 512 \div 1.575 = \sqrt[3]{325} = 6.875 \text{ inches.} \quad Ans.$$

WEIGHT OF HOLLOW BALLS OR SHELLS.

The weight of a hollow ball is the weight of a solid ball of the same diameter, *less* the weight of a solid ball whose diameter is that of the interior diameter of the shell.

EXAMPLE. — Required the weight of a cast iron shell whose exterior diameter is $6\frac{1}{4}$ inches, and interior diameter $4\frac{1}{4}$ inches.

$$6\tfrac{1}{4} = \tfrac{2.5}{1} \times \tfrac{2.5}{1} \times \tfrac{2.5}{1} = 244.14 \times .1365 = 33.33$$
$$4\tfrac{1}{4} = 4.25^3 \times .1365 \qquad\qquad\qquad = 10.48$$
$$\qquad\qquad\qquad\qquad\qquad\qquad\qquad 22.85 \text{ lbs.} \quad Ans.$$

Or, If we multiply the difference of the cubes, in inches, of the two diameters — the exterior and interior — by the weight of a spherical inch, we shall obtain the same result.

EXAMPLE. — Required the weight of a cast iron shell whose exterior diameter is 10 inches and interior diameter 8 inches.

$$\overline{10^3 - 8^3} \times .1365 = 66.612 \text{ lbs.} \quad Ans.$$

ANALYSIS OF COALS.

Description.	Volatile Matter.	Carbon.	Ash.
Breckinridge, Ky.,	62.25	29.10	8.65
" Albert," N. B.,	61.74	32.14	6.12
Chippenville, Pa.,	49.80		
Kanawha, "	41.85		
Pittsburg, "	32.95		
Cannel,	35.28	64.72	
Newcastle,	24.72	75.28	
Cumberland,	18.40	80.	1.60
Anthracite, a'v'g.,	3.43	89.46	7.11

Woods of most descriptions vary little from 80 per cent. volatile matter, and 20 per cent. charcoal.

TABLE—*Exhibiting the Weights, Evaporative Powers, &c., of Fuels, from Report of Professor Walter R. Johnson.*

Designation of Fuel.	Specific Gravity.	Weight per Cubic Foot.	Lbs. of Water at 212 degrees converted into Steam by 1 Cubic Foot of Fuel.	Lbs. of Water at 212 degrees converted into Steam by 1 lb. of Fuel.	Weight of Clinkers from 100 lbs. of Coal.
ANTHRACITE COALS.					
Beaver Meadow, No. 3	1.610	54.93	526.5	9.21	1.01
Beaver Meadow, No. 5	1.554	56.19	572.9	9.88	.60
Forest Improvement	1.477	53.66	577.3	10.06	.81
Lackawanna	1.421	48.89	493.0	9.79	1.24
Lehigh	1.590	55.32	515.4	8.93	1.08
Peach Mountain	1.464	53.79	581.3	10.11	3.03
BITUMINOUS COALS.					
Blossburgh	1.324	53.05	522.6	9.72	3.40
Cannelton, Ia.	1.273	47.65	360.0	7.34	1.64
Clover Hill	1.285	45.49	359.3	7.67	3.86
Cumberland, *average*,	1.325	53.60	552.8	10.07	3.33
Liverpool	1.262	47.88	411.2	7.84	1.86
Midlothian	1.294	54.04	461.6	8.29	8.82
Newcastle	1.257	50.82	453.9	8.66	3.14
Pictou	1.318	49.25	478.7	8.41	6.13
Pittsburgh	1.252	46.81	384.1	8.20	.94
Scotch	1.519	51.09	369.1	6.95	5.63
Sydney	1.338	47.44	386.1	7.99	2.25
COKE.					
Cumberland		31.57	284.0	8.99	3.55
Midlothian		32.70	282.5	8.63	10.51
Natural Virginia	1.323	46.64	407.9	8.47	5.31
WOOD.					
Dry Pine Wood		21.01	98.6	4.69	

MENSURATION OF LUMBER.

To find the contents of a board.

RULE. — Multiply the length in feet by the width in inches, and divide the product by 12; the quotient will be the contents in square feet.

EXAMPLE. — A board is 16 feet long and 10 inches wide; how many square feet does it contain?

$16 \times 10 = 160 \div 12 = 13\frac{4}{12}$. *Ans.*

To find the contents of a plank, joist, or stick of square timber.

RULE. — Multiply the product of the depth and width in inches by the length in feet, and divide the last product by 12; the quotient is the contents in feet, *board measure.*

EXAMPLE. — A joist is 16 feet long, 5 inches wide, and 2½ inches thick; how many feet does it contain, board measure?

$5 \times 2.5 \times 16 \div 12 = 16\frac{8}{12}$. *Ans.*

To find the solidity of a plank, joist, or stick of square timber.

RULE. — Multiply the product of the depth and width in inches by the length in feet, and divide the last product by 144; the quotient will be the contents in cubic feet.

EXAMPLE. — A stick of timber is 10 by 6 inches, and 14 feet in length; what is its solidity?

$10 \times 6 = 60 \times 14 = 840 \div 144 = 5\frac{5}{6}$ feet. *Ans.*

NOTE. — If a board, plank, or joist is narrower at one end than the other, add the two ends together and divide the sum by 2; the quotient will be the mean width. And if a stick of squared timber, whose solidity is required, is narrower at one end than the other $(A + a + \sqrt{Aa}) \div 3 =$ mean area. *A* and *a* being the areas of the ends.

To measure round timber.

RULE (IN GENERAL PRACTICE.) — Multiply the length, in feet, by the square of ¼ the girt, in inches, taken about ⅓ the distance from the larger end, and divide the product by 144; the quotient is considered the contents in cubic feet. For a strictly correct rule for measuring round timber, see MENSURATION OF SOLIDS — *Frustum of a Cone.*

EXAMPLE. — A stick of round timber is 40 feet in length, and girts 88 inches; what is its solidity?

$88 \div 4 = 22 \times 22 = 484 \times 40 = 19360 \div 144 = 134.44$ cub. ft. *Ans.*

MENSURATION OF LUMBER.

The following TABLE *is intended to facilitate the measuring of Round Timber, and is predicated upon the foregoing* RULE.

¼ Girt in Inches.	Area in Feet.	¼ Girt in Inches.	Area in Feet.	¼ Girt in Inches.	Area in Feet.	¼ Girt in Inches.	Area in Feet.
6	.25	12	1.	18	2.25	24	4.
6¼	.272	12¼	1.042	18¼	2.313	24¼	4.084
6½	.294	12½	1.085	18½	2.376	24½	4.168
6¾	.317	12¾	1.129	18¾	2.442	24¾	4.254
7	.34	13	1.174	19	2.506	25	4.34
7¼	.364	13¼	1.219	19¼	2.574	25¼	4.428
7½	.39	13½	1.265	19½	2.64	25½	4.516
7¾	.417	13¾	1.313	19¾	2.709	25¾	4.605
8	.444	14	1.361	20	2.777	26	4.694
8¼	.472	14¼	1.41	20¼	2.898	26¼	4.785
8½	.501	14½	1.46	20½	2.917	26½	4.876
8¾	.531	14¾	1.511	20¾	2.99	26¾	4.969
9	.562	15	1.562	21	3.062	27	5.062
9¼	.594	15¼	1.615	21¼	3.136	27¼	5.158
9½	.626	15½	1.668	21½	3.209	27½	5.252
9¾	.659	15¾	1.722	21¾	3.285	27¾	5.348
10	.694	16	1.777	22	3.362	28	5.444
10¼	.73	16¼	1.833	22¼	3.438	28¼	5.542
10½	.766	16½	1.89	22½	3.516	28½	5.64
10¾	.803	16¾	1.948	22¾	3.598	28¾	5.74
11	.84	17	2.006	23	3.673	29	5.84
11¼	.878	17¼	2.066	23¼	3.754	29¼	5.941
11½	.918	17½	2.126	23½	3.835	29½	6.044
11¾	.959	17¾	2.187	23¾	3.917	30	6.25

To find the solidity of a log by help of the preceding TABLE.

RULE.—Multiply the tabular area opposite the corresponding ¼ girt, by the length of the log in feet, and the product will be the solidity in feet.

EXAMPLE.—The ¼ girt of a log is 22 inches, and the length of the log is 40 feet; required the solidity of the log.

3.362 × 40 = 134.48 cubic feet. *Ans.*

NOTE.—Though custom has established, in a very general way, the preceding method as that whereby to measure round timber, and holds, in most instances, the solidity to be that which the method will give, there seems, if the object sought be the real solidity of the stick, neither accuracy, justice, nor certainty, in the practice.

Thus, in the preceding example, the stick was supposed to be 40 feet in length, and 88 inches in circumference at ¼ the distance from the larger end, and was found, by the method, to contain 134.44 cubic feet: now 88 ÷ 3.1416 = 28 inches, = the diameter at ¼ the distance from the greater base, and retaining this diameter and the length, we may

58 MENSURATION OF LUMBER.

suppose, with sufficient liberality, and without being far from the general run of such sticks, the diameter at the greater base to be 30 inches, and that of the less to be 24 inches, and —

By a correct rule the stick contains —

$30 \times 24 = 720 + 12 = 732 \times .7854 \times 40 = 22996 \div 144 = 159.7$ cubic feet, or 19 per cent. more than given by the method under consideration; and we need hardly add that the nearer the stick approaches to the figure of a cylinder, the wider will be the difference between the truth and the result obtained by the method referred to. Thus, suppose the stick a cylinder, 28 inches in diameter, and 40 feet in length; and we have, by the fallacious rule, as above, 134.44 cubic feet; and —

By a correct method, we have —

$28^2 \times .7854 \times 40 = 24630 \div 144 = 171$ cubic feet, or over 27 per cent. more than furnished by the erroneous mode of practice.

Again: suppose the stick in the form of a cone, 30 inches at the base, and tapering to a point at 150 feet in length; and we have, by a correct rule —

$30^2 \div 3 = 300 \times .7854 \times 150 = 35343 \div 144 = 245.44$ cubic feet; and by the ordinary method of gauging, or the aforementioned practice, we have —

$20 \times 3.1416 = 62.832 \div 4 = 15.708^2 \times 150 = 37011.19 \div 144 = 257$ cubic feet, or nearly 4¾ per cent. more than the stick actually contains.

In short, without taking into account anything for the thickness of the bark, that may be supposed to be on the stick, the method is correct only when the stick tapers at the rate of 5¼ inches diameter per each 10 feet in length, or over ½ inch diameter to each foot in length of the stick.

If, however, we suppose the stick as before, (30 inches at the greater base, 24 inches at the smaller, and 40 feet in length,) and suppose the bark upon it to be 1 inch thick, we shall have, by the usual method, 134.44 cubic feet, as before. And, exclusive of the bark, by a correct method, we shall have.

$\overline{30-2} \times \overline{24-2} = 616 + 12 = 628 \times .7854 \times 40 = 19729 \div 144 = 137$ cubic feet, or only about 2 per cent. more than that furnished us by the usual practice.

The following simple rule for measuring round timber is sufficiently correct for most practical purposes: —

RULE. — Multiply the square of one-fifth of the mean girt, (exclusive of bark,) in inches, by twice the length of the stick in feet, and divide the product by 144; the quotient will be the solidity in feet.

To find the solidity of the greatest rectangular stick that may be cut from a given log, or from a stick of round timber of given dimensions.

RULE. — Multiply the square of the mean diameter of the log, in inches, by half the length of the log, in feet, and divide the product by 144.

EXAMPLE. — The diameter (exclusive of bark) of the greater base of a stick of round timber is 30 inches, and that of the less base is 24 inches, and the stick is 40 feet in length; required the solidity of the greatest rectangular stick that may be cut from it.

$30 \times 24 + \frac{1}{3}(30-24)^2 = 732 =$ square of mean diameter,* and

$732 \times 20 = 14640 \div 144 = 101\frac{2}{3}$ cubic feet. *Ans.*

---

* Except in the case of a cylinder, there is a difference betwixt the *mean* diameter of a solid having circular bases, and the *middle* diameter of that solid. The mean diameter reduces the solid to a cylinder; the middle diameter is the diameter midway between the two bases.

MENSURATION OF LUMBER. 59

Note. — The foregoing stick will make —
14640 ÷ 16 = 915 feet of square-edged boards 1 inch thick ;
Or, 101⅔ × 9 = 915.

To find the solidity of the greatest square stick that may be cut from a given log, or from a stick of round timber of given dimensions.

RULE. — Multiply the square of the diameter of the less end of the log, in inches, by half the length of the log, in feet, and divide the product by 144.

EXAMPLE. — The preceding supposed log will make a square stick containing —

$$24^2 \times \tfrac{40}{2} = 1152 \div 144 = 80 \text{ cubic feet.}$$

Diameter multiplied by .7071 = side of inscribed square.

To find the contents, in Board Measure, of a log, no allowance being made for wane or saw-chip.

RULE. — Multiply the square of the mean diameter, in inches, by the length in feet, and divide the product by 15.28.

Or, Multiply the square of the mean diameter in inches, by the length in feet, and that product by .7854, and divide the last product by 12.

The cubic contents of a log multiplied by 12, equal the contents of the log, board measure.

The convex surface of a Frustum of a Cone = $(C + c) \times \tfrac{1}{2}$ slant length; C being the circumference of the greater base, and c the circumference of the less.

GAUGING.

Rules *for finding the capacity in gallons or bushels of different shaped Cisterns, Bins, Casks, &c., and also, by way of examples, for constructing them to given capacities.*

Rule — 1. *When the vessel is rectangular.* Multiply the interior length, breadth, and depth, in feet together, and the product by the capacity of a cubic foot, in gallons or bushels, as desired for its capacity.

Rule — 2. *When the vessel is cylindrical.* Multiply the square of its interior diameter in feet, by its interior depth in feet, and the product by the capacity of a *cylindrical* foot in gallons or bushels, as desired for its capacity.

Rule — 3. *When the vessel is a rhombus or rhomboid.* Multiply its interior length, in feet, its right-angular breath in feet, and its depth in feet together, and the product by the capacity of a cubic foot in the special measure desired for its capacity.

Rule — 4. *When the vessel is a frustum of a cone* — a round vessel larger at one end than the other, whose bases are planes. Multiply the interior diameter of the two ends together, in feet, add ⅓ the square of their difference in feet to the product, multiply the sum by the perpendicular depth of the vessel in feet, and that product by the capacity of a *cylindrical* foot in the unit of measure desired for its capacity.

Rule — 5. *When the vessel is a prismoid or the frustum of any regular pyramid.* To the square root of the product of the areas of its ends in feet, add the areas of its ends in feet, multiply the sum by ⅓ its perpendicular depth in feet, and that product by the capacity of a cubic foot in gallons or bushels, as desired for its capacity.

If it is found more convenient to take the dimensions in inches, do so; proceed as directed for feet, divide the product by 1728, and multiply the quotient by the capacity of the respective foot as directed. Or, multiply the capacity in inches by the capacity of the respective inch in gallons or bushels; — by the quotient obtained by dividing the capacity of the respective foot in gallons or bushels by 1728 — for the contents.

Rule — 6. *When the vessel is a barrel, hogshead, pipe, &c.* Multiply the difference in inches between the bung diameter and head diameter, (interior,) if the staves be

much curved, . . by .7
medium curved, . by .65   } *See* page 63.
straighter than medium, by .6
nearly straight, . by .55

and add the product to the head diameter, taken in inches; then multiply the square of the sum by the length of the cask in inches, and divide the product by the capacity in *cylindrical* inches of a gallon or

GAUGING.

bushel as desired for the contents. Or, divide the contents in *cylindrical* inches, as above found, by 1728, and multiply the quotient by the capacity of a cylindrical foot in gallons or bushels as desired for its contents. Or, multiply the capacity in cylindrical inches by the capacity of a cylindrical inch, in gallons or bushels, as desired, — that is, by the quotient obtained by dividing the capacity of a cylindrical foot in gallons or bushels, by 1728, for the contents.

The capacity of a

CUBIC FOOT =		CYLINDRICAL FOOT =	
7.4805	Winchester wine gallons.	5.8751	Winchester wine gallons.
6.1276	Ale "	4.8126	Ale "
6.2321	Imperial "	4.8947	Imperial "
.80356	Winchester bushel.	.63111	Winchester bushel.
.62888	" heaped "	.49391	" heaped "
.64285	" 1¼ even "	.50489	" 1¼ even "
.779	Imperial " "	.61183	Imperial "

EXAMPLE. — Required the capacity in Winchester bushels of a rectangular bin, whose interior length is 12 feet, breadth 6 feet, and depth 5 feet.

$12 \times 6 \times 5 \times .8035 = 289.26$ bushels. *Ans.*

EXAMPLE. — Required the capacity in Winchester wine gallons of a cylindrical can, whose interior diameter is 18 inches, and depth 3 feet.

$18 \times 18 \times 36 \times 5.875 \div 1728 = 39.66$ gallons. *Ans.*

Or, $1.5 \times 1.5 \times 3 \times 5.875 \quad = 39.66$ gallons. *Ans.*

Or, $18 \times 18 \times 36 \times .0034 \quad = 39.66$ gallons. *Ans.*

EXAMPLE. — How many Winchester bushels in 39.66 wine gallons?

$39.66 \times .10742 = 4.26$ bushels. *Ans.*

EXAMPLE. — How many wine gallons in 4.26 Winchester bushels?
$4.26 \times 9.3092 = 39.66$ gallons. *Ans.*

EXAMPLE. — How many wine gallons will a cistern in the form of a frustum of a cone hold, having the interior diameter of one of its ends 6 feet, and that at the other 8 feet, and its perpendicular depth 9 feet?

$8 - 6 = 2$, and $2^2 \div 3 = 1.333 = \frac{1}{3}$ square of dif. of diameters, and
$\quad 6 \times 8 + 1.333 = 49.333 \times 9 \times 5.8751 = 2608.55$ gals. *Ans.*

Or, $6 \times 8 + 8^2 + 6^2 = 148 \times \frac{2}{3} \times 5.8751 = 2608.55$ gals. *Ans.*

Or, $(8^3 - 6^3) \div (8 - 6) = 148 \times \frac{2}{3} \times 5.8751 = 2608.55$ gals. *Ans.*

Or, $96 - 72 = 24$ and $(24^2 \div 3) = 192$, and
$\quad 96 \times 72 + 192 = 7104 \times 108 \times .0034 = 2608.55$ gals. *Ans.*

GAUGING.

Example. — What is the capacity in Winchester bushels of a cistern whose form is prismoid, the dimensions (interior) of one end being 8 by 6 feet, of the other 4 by 3 feet, and its perpendicular depth 12 feet?

$8 \times 6 = 48 =$ area of one end, and $4 \times 3 = 12 =$ area of the other end; then —

$48 \times 12 = \sqrt{576} = 24 + 48 + 12 = 84 \times \frac{12}{3} \times .80356 = 270$ bushels. *Ans.*

Or, $(8 + 4) \div 2 = 6$, and $(6 + 3) \div 2 = 4.5 =$ mean sectional areas of ends, and
$6 \times 4.5 \times 4 = 4$ area of mean perimeter, then
$8 \times 6 + 4 \times 3 + 6 \times 4.5 \times 4 = 168 \times \frac{12}{6} \times .80356 = 270$ bus. *Ans.*

Example. — What must be the depth of a rectangular bin whose length is 12 feet, and breadth 6 feet, to hold 289.26 bushels?

$289.26 \div (12 \times 6 \times .80356) = 5$ feet. *Ans.*

Example. — A cylindrical can, whose depth is to be 36 inches, is required to be made that will hold 40 gallons; what must be the diameter of the can?

$40 \div (3 \times 5.8751) = \sqrt{2.27} = 1.506$ feet. *Ans.*
Or, $40 \div (36 \times .0034) = \sqrt{326.8} = 18.07$ inches. *Ans.*

Example. — A cylindrical can, whose interior diameter is to be 18 inches, is required that will hold 40 gallons; what must be the interior depth of the can?

$40 \div (18^2 \times .0034) = 36.31$ inches. *Ans.*
Or, $40 \div (1.5^2 \times 5.8751) = 3.026$ feet. *Ans.*

Example. — A cistern is to be built in the form of a frustum of a cone, that will hold 1800 gallons, and the diameter of one of its ends is to be 5 feet, and that of the other 7½ feet; what must be the depth?

$7.5 - 5 = 2.5$, and $2.5^2 \div 3 = 2.0833 = \frac{1}{3}$ square of difference of diameter, and

$1800 \div (7.5 \times 5 + 2.0833) \times 5.8751 = 7.74$ feet. *Ans.*

Or, $1800 \div \left(\dfrac{7.5 \times 5 + 7.5^2 + 5^2}{3} \times 5.8751 \right) = 7.74$ feet. *Ans.*

Example. — The form, capacity, depth, and diameter of one end being determined on, and being as above, what must be the diameter of the other end?

$\dfrac{c}{\frac{1}{3}h} - \frac{1}{3}d^2 = y$, c being the solidity in cylindrical measurement, h

the depth, d the diameter of the given end or base, and y a quantity the square root of which is the sum of the required base and half the given base; then

$1800 \div 5.8751 = 306.378 =$ solidity in cylindrical feet, and

$306.378 \div \frac{7.74}{2} = 118.75 - (5^2 \div \frac{4}{3}) = \sqrt{100} = 10 - \frac{5}{2} = 7.5$ feet. *Ans.*

EXAMPLE. — A measure is to be built in the form of a frustum of a cone, that will hold exactly 1 wine gallon, and the diameter of one of its ends is to be 4 inches, and that of the other 6 inches; what must be its depth?

$1 \div (6 \times 4 + 1\frac{1}{3}) \times .0034 = 11.61$ inches. *Ans.*

Or, $\dfrac{231}{.7854} \div \dfrac{6 \times 4 + 6^2 + 4^2}{3} = 11.61$ inches. *Ans.*

EXAMPLE. — A measure in the form of a frustum of a cone holds 1 wine gallon; the diameter of one of its ends is 6 inches, and its depth is 11.61 inches; what is the diameter of the other end?

$\dfrac{231}{.7854} = 294.1176 \div \dfrac{11.61}{3} = 76 - (6^2 \div \frac{4}{3}) = \sqrt{49} = 7 - \frac{6}{2} = 4$ inches. *Ans.*

CASK GAUGING.

CASK-GAUGING, in a general sense, is a practical art, rather than a scientific achievement or problem, and makes no pretensions to strict accuracy with regard to the conclusions arrived at. The aim is, by means of a few satisfactory measurements taken of the outside, and an estimate of the probable mean thickness of the material of which the cask is composed (of which there must always remain some doubt), or by means of a few measurements taken of the inside, to determine, 1st, the capacity of the cask, and, 2d, the ullage, or capacity of the occupied or unoccupied space in a cask but partly full. And the Rule (RULE 6, page 60), which reduces the supposed cask, or cask of supposed curvature, to a cylinder, is as practically correct for the capacity of ordinary casks, as any rule, or set of rules, that can be offered for general purposes.

Casks have no fixed form of their own, to which they severally and collectively correspond, nor are they in any considerable degree in conformity with any regular geometrical figure.

Some casks — a few — those having their staves much curved throughout their entire length, are nearest in keeping with the *middle frustum of a spheroid;* others, slightly less curved than the preceding, correspond in a considerable degree to the *middle frustum of a parabolic spindle;* others, again — those having very little longitudinal curvature of stave to their semi-lengths — are nearly in keeping with the *equal frustums of a paraboloid;* and others — a very few — those whose staves are straight from the bung diameter to the heads, or equal to that form, are in accordance with the *equal frustums of a cone.*

The *gauging rod*, which is intended to be correct for casks of the most common form, gives for all casks, as may be seen in one of the following EXAMPLES, a solidity slightly greater (about 2½ per cent.) than would be obtained by supposing the cask in conformity with the third figure above alluded to.

The RULE for finding the contents of a cask, *by four dimensions*, hereafter to be given, is intended as a general Rule for all casks, and, when the diameter midway between the bung and head can be accurately ascertained, will lead to a very close approach to the truth.

From the length of a cask, taken from outside to outside of the heads, with callipers, it is usual to deduct from 1 to 2 inches, to correspond with the thickness of the heads, according to the size of the cask, and the remainder is taken as the length of the interior.

To the diameter of each head, taken externally, from ¼ inch to $\frac{3}{10}$ inch should be added for common-sized barrels, $\frac{4}{10}$ inch for 40 gallon casks, and from ½ inch to $\frac{6}{10}$ inch for larger casks, to correspond with the interior diameters of the heads.

If the staves are of uniform thickness, any sectional diameter of a cask may be nearly or quite ascertained, by dividing the circumference at that place by 3.1416, and subtracting twice the thickness of the stave from the quotient.

For obtaining the diagonal of a cask by mathematical process, — the interior length, &c. &c. — see *Rules*, below.

In the following formulas D denotes the bung diameter, d the head diameter, and l the length of the cask.

The solidity of any cask is equal to its length multiplied by the square of its mean diameter multiplied by .7854.

To calculate the contents of a cask from four dimensions.

RULE. — To the square of the bung diameter add the square of the head diameter, and the square of double the diameter midway between the bung and head, and multiply the sum by ⅙ the length of the cask, for its *cylindrical* contents; the product multiplied by .0034 expresses the contents in wine gallons.

EXAMPLE. — The length of the cask is 40 inches, its bung diameter 28 inches, head diameter 20 inches, and the diameter midway between the bung and head is 25.6 inches; how many gallons' capacity has the cask?

$20^2 + 28^2 + \overline{25.6 \times 2}^2 = 3805.44 \times \frac{4.0}{6} \times .0034 = 86.26$ gals. *Ans.*

$(D^2 + d^2 + \overline{2m}^2) \times \frac{1}{6} l \times .7854 =$ cubic contents.

$\dfrac{D^2 + d^2 + \overline{2m}^2}{6} =$ square of mean diameter.

By RULE 6, p. 68, this cask will hold —

$28 - 20 = 8 \times .65 = 5.2 + 20 = 25.2 \times 25.2 \times 40 \times .0034 = 86.36$ gallons.

When the cask is in the form of the middle frustum of a spheroid.

$\frac{2}{3} D^2 + \frac{1}{3} d^2 =$ square of mean diameter.

And a cask of this form, having the same head diameter, bung diameter, and length as the preceding, will hold —

$\dfrac{2 \times 28^2 + 20^2}{3} \times 40 \times .0034 = 89.216$ gallons.

When the cask is in the form of the middle frustum of a parabolic spindle.

$\frac{2}{3} D^2 + \frac{1}{3} d^2 - \frac{2}{15} (D \frown d)^2 =$ square of mean diameter.

And a cask of this form, having the same head diameter, bung diameter, and length as the preceding, will hold —

$522\frac{2}{3} + 133\frac{1}{3} = 656 - 8.533 = 647.467 \times 40 \times .0034 = 88.055$ gals.

When the cask is in the form of two equal frustums of a paraboloid.

$\frac{1}{2} D^2 + \frac{1}{2} d^2 =$ square of mean diameter.

And a cask of this form, having the same head diameter, bung diameter, and length as the preceding, will hold —

$\dfrac{28^2 + 20^2}{2} \times 40 \times .0034 = 80.51$ gallons.

When the cask is in the form of the equal frustums of a cone.

$\frac{1}{2} D^2 + \frac{1}{2} d^2 - \frac{1}{6} (D \frown d)^2 =$ square of mean diameter.

Or, $\frac{1}{3} D^2 + \frac{1}{3} d^2 + \frac{1}{3} Dd =$ " " " "

Or, $D \times d + \frac{1}{3} (D \frown d)^2 =$ " " " "

And a cask of this form, having the same head diameter, bung diameter, and length as the preceding, will hold —

$\overline{28 \times 20 + 21\frac{1}{3}} \times 40 \times .0034 = 79.06$ gals.

To find the contents of a cask the same as would be given by the gauging rod.

The *gauging rod* is constructed upon the principle that the cube of the diagonal of a cask, in inches, multiplied by $\frac{1}{61050}$, equals the contents of the cask, in Imperial gallons.

The contents in wine gallons of either of the aforementioned casks, therefore, by the gauging rod, would be —

$$\overline{31.241}^3 \times .0027 = 82\tfrac{1}{4} \text{ gals.}$$

The decimal coëfficient to take the place of .0027, for finding the contents of a cask in the form of the middle frustum of a spheroid = .002926; and for finding the contents of a cask in the form of the equal frustums of a cone = .002593. And between these extremes lies the decimal for other casks, or casks of intervening figures.

To find the diagonal of a cask, when the interior is inaccessible.

RULE. — From the bung diameter subtract half the difference of the bung and head diameters, and to the square of the remainder add the square of half the length of the cask, and the square root of the sum will be the diagonal.

EXAMPLE. — What is the diagonal of a cask whose bung diameter is 28 inches, head diameter 20 inches, and length 40 inches?

$$28 - 20 = 8 \div 2 = 4, \text{ and } 28 - 4 = 24, \text{ then}$$
$$\sqrt{(24^2 + 20^2)} = 31.241 \text{ inches. } Ans.$$

To find the length of a cask, the head diameter, bung diameter and diagonal being given.

$$\sqrt{\left(\text{diagonal}^2 - \overline{D - \frac{D \backsim d}{2}}^2\right)} = \tfrac{1}{2} l.$$

And the interior length of a cask, whose interior head diameter, bung diameter and diagonal, are as the preceding, will be

$$\sqrt{(31.241^2 - 24^2)} = 20 \times 2 = 40 \text{ inches.}$$

To find the solidity of a sphere.

$D^2 \times \tfrac{2}{3} D \times .7854 =$ cubic contents, D being the diameter.

To find the solidity of a spherical frustum.

$$\left(\tfrac{2}{3} h^2 + \frac{b^2 + d^2}{2}\right) \times h \times .7854 = \text{cubic contents, } b \text{ and } d \text{ being the bases, and } h \text{ the height.}$$

NOTE. — For Rules in detail pertaining to the foregoing figures, and for other figures, see MENSURATION OF SOLIDS.

ULLAGE.

The *ullage* or *wantage* of a cask is the quantity the cask lacks of being full.

To find the ullage of a standing cask, when the cask is half full or more.

RULE. — To the square of the head diameter, add the square of the diameter at the surface of the liquor, and the square of twice the diameter midway between the surface of the liquor and the upper head, and divide the sum by 6; the quotient, multiplied by the distance from the surface of the liquor to the upper head, multiplied by .0034, will give the ullage in wine gallons.

EXAMPLE. — The diameters are as follows — at the upper head, 20 inches; at the surface of the liquor, 22 inches; and at a point midway between these, $21\frac{1}{4}$ inches; and the distance from the upper head to the surface of the liquor is 5 inches; required the ullage.

($20^2 + 22^2 + \overline{21.25 \times 2}^2$) $\div$ 6 = 448.37 $\times$ 5 $\times$ .0034 = 7.62 gallons. *Ans.*

When the cask is standing, and less than half full, to find the ullage.

RULE. — Make use of the bung diameter in place of the head diameter, and proceed in all respects as directed in the last *Rule*, and add the quantity found to half the capacity of the cask; the sum will be the ullage.

EXAMPLE. — The bung diameter is 28 inches; the diameter at the surface of the liquor, below the bung, is 26 inches; the diameter midway between the bung and the surface of the liquor is 27.3 inches; and the distance from the surface of the liquor to the bung diameter is 5 inches; required the quantity the cask lacks of being half full; and also the ullage of the cask, its capacity being 86.26 gallons.

($28^2 + 26^2 + \overline{27.3 \times 2}^2$) $\div$ 6 = 740.2 $\times$ 5 $\times$ .0034 = 12.58 gallons less than $\frac{1}{2}$ full. *Ans.*

And, 86.26 $\div$ 2 = 43.13 + 12.58 = 55.73 gallons ullage. *Ans.*

When the cask is upon its bilge, and half full or more, to find the ullage.

RULE. — Divide the distance from the bung to the surface of the liquor — (the height of the empty segment) — by the whole bung diameter, and take the quotient as the height of the segment of a circle whose diameter is 1, and find the area of the segment; multiply the area by the capacity of the cask, in gallons, and that product by 1.25; the last product will be the ullage, in gallons, as found by the aid of the *wantage-rod;* and will be correct for casks of the most common form.

NOTE. — The area of the segment of a circle =
(ch'd $\frac{1}{2}$ arc $+$ $\frac{1}{3}$ ch'd $\frac{1}{2}$ arc $+$ ch'd $\frac{1}{2}$ seg.) $\times$ height seg. $\times \frac{4}{10}*$, very nearly.
And, having the diameter of the circle and the height of the segment given, the chord of half the arc, and the chord of the segment may be found, thus —
radius — height = cosine; radius^2 — cosine^2 = sine^2; $\sqrt{}$ (sine^2) $\times 2$ = ch'd of seg.
sine^2 + height seg.^{2} = ch'd $\frac{1}{2}$ arc^2, and $\sqrt{}$ *(ch'd $\frac{1}{2}$ arc2) = ch'd $\frac{1}{2}$ arc.*

EXAMPLE. — The bung diameter is 28 inches, the height of the empty segment 5.6 inches, and the capacity of the cask 86.26 gallons; required the ullage of the cask, in gallons.

$5.6 \div 28 = .2 =$ height of seg., diameter as 1.
$1 \div 2 = .5 =$ radius.
$.5 - .2 = .3 =$ cosine.
$.5^2 - .3^2 = .16 =$ sine^2, or square of half the base of the segment.
$\sqrt{.16} = .4 \times 2 = .8 =$ chord of segment, or base of segment.
$.4^2 + .2^2 = .2 =$ square of chord of half the arc.
$\sqrt{.2} = .4472 =$ chord of half the arc, then —
$.4472 \div 3 = .1491$, and $\overline{.1491 + .4472 + .8} \times .2 \times \frac{4}{10} = .1117$, area of segment, and
$.1117 \times 86.26 \times 1.25 = 12$ gallons. *Ans.*

When the cask is upon its bilge, and less than half full, to find the ullage.

RULE. — Divide the depth of the liquor by the bung diameter, and proceed in all respects as directed in the last Rule; then subtract the quantity found from the capacity of the cask, and the difference will be the ullage of the cask.

To find the quantity of liquor in a cask by its weight.

EXAMPLE. — The weight of a cask of proof spirits is 300 lbs., and the weight of the empty cask (*tare*) is 32 lbs. How many gallons are there of the liquor?

$300 - 32 = 268 \div 7.732 = 34\frac{2}{3}$ gallons. *Ans.*

Customary Rule by Freighting Merchants, for finding the cubic measurement of casks.

Bung diameter$^2 \times \frac{2}{3}$ length of cask = cubic measurement.

NOTE. — One cubic foot contains 7.4805 wine gallons.

* For several Rules in detail, for finding the area of the segment of a circle, see GEOMETRY — *Mensuration of Superficies.*

TONNAGE.

GOVERNMENT MEASUREMENT.

$$\frac{\text{length} - \tfrac{3}{5}\text{ breadth} \times \text{breadth} \times \text{depth}}{95} = \text{tonnage}.$$

In a double-decked vessel, the length is reckoned from the fore part of the main stem to the after side of the sternpost above the upper deck; the breadth is taken at the broadest part above the main wales, and half this breadth is taken for the depth.

In a single-decked vessel the length and breadth are taken as for a double-decked vessel, and the distance between the ceiling of the hold and the under side of the deck plank is taken as the depth.

EXAMPLE. — The length of a double-decked vessel is 260 feet, and the breadth is 60 feet; required the tonnage.

$260 - \frac{60 \times 3}{5} = 224 \times 60 \times \frac{60}{2} = 403200 \div 95 = 4244.2$ tons. *Ans.*

EXAMPLE. — The length of a single-decked vessel is 180 feet, the breadth 34 feet, and depth 18 feet; required the tonnage.

$180 - \tfrac{3}{5}$ of $34 = 159.6 \times 34 \times 18 \div 95 = 1028.16$ tons. *Ans.*

CARPENTER'S MEASUREMENT.

For a double-decked —

$$\frac{\text{length of keel} \times \text{breadth main beam} \times \tfrac{1}{2}\text{ breadth}}{95} = \text{tonnage}.$$

For a single-decked —

$$\frac{\text{length of keel} \times \text{breadth main beam} \times \text{depth of hold}}{95} = \text{tonnage}.$$

OF CONDUITS OR PIPES.

Pressure of Water in Vertical Pipes, &c.

h = height of column in inches ; o = circumference of column in inches; t = thickness of pipe in inches equal in strength to lateral pressure at base of column ; w = weight of a cubic inch of water in pounds ; C = cohesive strength in pounds per inch area of transverse section of the material of which the pipe is composed — TABLE, p. 72.

ho = area of interior of pipe in inches ; hw = pressure in pounds per square inch at the base of the column, or maximum lateral pressure in pounds per square inch on the pipe tending to burst it ; how = maximum lateral pressure in pounds on the pipe, tending to burst it at the bottom ; and $how \div 2$ = mean lateral pressure in pounds on the pipe, or pressure in pounds on the pipe tending to burst it at half the height of the column.

$$how \div C = t; \quad how \div t = C; \quad Ct \div ow = h; \quad Ct \div hw = o.$$

NOTE. — The *reliable* cohesion of a material is not above ¼ its ultimate force, as given in the Table of Cohesive Forces. By experiment, it has been found that a cast iron pipe 15 inches in diameter and ¼ of an inch thick, will support a head of water of 600 feet ; and that one of the same diameter made of oak, and two inches thick, will support a head of 160 feet : 12000 lbs. per square inch for cast iron, 1200 for oak, 750 for lead, are counted safe estimates. The ultimate cohesion of an alloy, composed of lead 8 parts and zinc 1 part, is 3000 pounds per square inch.

Concerning the Discharge of Pipes, &c.

Small pipes, whether vertical, horizontal, or inclined, under equal heads, discharge proportionally less water than large ones. That form of pipe, therefore, which presents the least perimeter to its area, other things being equal, will give the greatest discharge. A round pipe, consequently, will discharge more water in a given time than a pipe of any other form, of equal area.

The greater the length of a pipe discharging vertically, the greater the discharge. Because the friction of the particles against its sides, and consequent retardation, is more than overcome by the gravity of the fluid.

The greater the length of a pipe discharging horizontally, the less proportionally will be the discharge. The proportion compared with a less length is in the inverse ratio of the square root of the two lengths, nearly.

Other things being equal, rectilinear pipes give a greater discharge than curvilinear, and curvilinear greater than angular. The head, the diameters and the lengths being the same, the time occupied in passing an equal quantity of water through a straight pipe is 9, through one curved to a semicircle 10, and through one having one right angle, otherwise straight, 14. All interior inequalities and roughness should be avoided.

It has been ascertained that a velocity of 60 feet a minute (1 foot a second) through a horizontal pipe, 4 inches in diameter and 100 feet

CONDUITS OR PIPES.   71

in length, is produced by a head $2\frac{1}{4}$ inches, only $\frac{1}{4}$ of an inch above the upper surface of the orifice; and that, to maintain an equal velocity through a pipe similarly situated, of equal length, having a diameter of $\frac{1}{2}$ inch only, a head of $1\frac{5}{12}$ feet is required. To increase the velocity through the last mentioned pipe to 2 feet a second, requires a head $4\frac{9}{12}$ feet; to 3 feet, a head of $10\frac{1}{12}$; to 4 feet, a head of $17\frac{9}{12}$, &c.

From the foregoing, the following, it is believed, reliable rules, are deduced.

To find the velocity of water passing through a straight horizontal pipe of any length and diameter, the head, or height of the fluid above the centre of the orifice, being known.

RULE. — Multiply the head, in feet, by 2500, and divide the product by the length of the pipe, in feet, multiplied by 13.9, divided by the interior diameter of the pipe in inches; the square root of the quotient will be the velocity in feet per second.

EXAMPLE. — The head is 6 feet, the length of the pipe 1340 feet, and its diameter 5 inches; required the velocity of the water passing through it.

$2500 \times 6 = 15000 \div (\frac{1340 \times 13.9}{5}) = \sqrt{4.03} = 2$ feet per second. *Ans.*

To find the head necessary to produce a required velocity through a pipe of given length and diameter.

RULE. — Multiply the square of the required velocity, in feet, per second, by the length of the pipe multiplied by the quotient obtained by dividing 13.9 by the diameter of the pipe in inches, and divide the product thus obtained by 2500; the quotient will be the head in feet.

EXAMPLE. — The length of a pipe lying horizontal and straight is 1340 feet, and its diameter is 5 inches; what head is necessary to cause the water to flow through it at the rate of 2 feet a second?

$2^2 \times 1340 \times \frac{13.9}{5} \div 2500 = 6$ feet. *Ans.*

To find the quantity of water flowing through a pipe of any length and diameter.

RULE. — Multiply the velocity in feet per second by the area of the discharging orifice, in feet, and the product is the quantity in cubic feet discharged per second.

EXAMPLE. — The velocity is 2 feet a second, and the diameter of the pipe 5 inches; what quantity of water is discharged in each second of time?

$5 \div 12 = .4166$, and $.4166^2 \times .7854 \times 2 = .273$ cubic foot. *Ans.*

MISCELLANEOUS PROBLEMS.

To find the specific gravity of a body heavier than water.

RULE. — Weigh the body in water and out of water, and divide the weight out of water by the difference of the two weights.

EXAMPLE. — A piece of metal weighs 10 lbs. in atmosphere, and but 8¼ in water; required its specific gravity.

$$10 - 8.25 = 1.75, \text{ and } 10 \div 1.75 = 5.714. \quad Ans.$$

To find the specific gravity of a body lighter than water.

RULE. — Weigh the body in air; then connect it with a piece of metal whose weight, both in and out of water, is known, and of sufficient weight that the two will sink in water; and find their combined weight in water; then divide the weight of the body in air by the weight of the two substances in air, less the sum of the difference of the weight of the metal in air and water and the combined weight of the two substances in water, and the quotient will be the specific gravity sought.

EXAMPLE. — The combined weight, in water, of a piece of wood, and piece of metal, is 4 lbs.; the wood weighs in atmosphere 10 lbs.; and the metal in atmosphere 12, and in water 11 lbs.; required the specific gravity of the wood.

$$10 \div (10 + 12 - \overline{12 \backsim 11 + 4}) = .588. \quad Ans.$$

To find the specific gravity of a fluid.

RULE. — Multiply the known specific gravity of a body by the difference of its weight in and out of the fluid, and divide the product by its weight out of the fluid; the quotient will be the specific gravity of the fluid in which the body is weighed.

EXAMPLE. — The specific gravity of a brass ball is 8.6; its weight in atmosphere is 8 oz., and in a certain fluid 7¼ oz.; required the specific gravity of the fluid.

$$8 - 7.25 = .75, \text{ and } 8.6 \times .75 = 6.45, \text{ and } 6.45 \div 8 = .806. \quad Ans.$$

To find the proportion of one to the other of two simples forming a compound, or the extent to which a metal is debased, (the metal and the alloy used being known.)

The Rule strictly bears upon that of *Alligation Alternate,* which see.

EXAMPLE. — The specific gravity of gold is 19.258, and that of copper, 8.788; an article composed of the two metals, has a specific gravity of 18; in what proportion are the metals mixed?

$$18 \backsim 19.258 \times 8.788 = 11.055$$
$$18 \backsim 8.788 \times 19.258 = 177.4, \text{ then}$$

MISCELLANEOUS PROBLEMS. 73

$$\frac{11.055 + 177.4}{11.055 + 177.4} : 11.055 :: 18 = 1.056 \text{ copper,}$$
$$\frac{11.055 + 177.4}{11.055 + 177.4} : 177.4 :: 18 = 16.944 \text{ gold.}$$ } Ans.

Or, 18 — 1.056 = 16.944 gold. Copper to gold as 1 to 16.04 +

To find the lifting power of a balloon.

RULE. — Multiply the capacity of the balloon, in feet, by the difference of weight between a cubic foot of atmosphere and a cubic foot of the gas used to inflate the balloon, and the product is the weight the balloon will raise.

EXAMPLE. — A balloon, whose diameter is 24 feet, is inflated with hydrogen ; what weight will it raise ?

Specific gravity of air is 1, weight of a cubic foot 527.04 grains; specific gravity of hydrogen is .0689.
527.04 × .0689 = 36.31 grains = weight of 1 cubic foot of hydrogen.
527.04 — 36.31 = 490.73 grs. = dif. of weight of air and hydrogen.
24^3 × .5236 = 7238.24 = capacity in cubic feet of balloon.
Then, 7238.24 × 490.73 = 3552021 grs. = $\frac{3552021}{7000}$ = 507$\frac{4}{10}$ lbs. Ans.

To find the diameter of a balloon that shall be equal to the raising of a given weight.

The weight to be raised is 507$\frac{4}{10}$ lbs.

$\overline{507.4 × 7000 ÷ 490.73}$ = 7238.24, and 7238.24 ÷ .5236 = $\sqrt[3]{13824}$ = 24 feet. Ans.

To find the thickness of a concave or hollow metallic ball or globe, that shall have a given buoyancy in a given liquid.

EXAMPLE. — A concave globe is to be made of brass, specific gravity 8.6, and its diameter is to be 12 inches; what must be its thickness that it may sink exactly to its centre in pure water ?

Weight of a cubic inch of water .036169 lb. ; of the brass .3112 lb.
Then, 12^3 × .5236 × .036169 ÷ 2 = 16.3625 cubic inches of water to be displaced.
16.3625 ÷ .3112 = 52.5787 cubic inches of metal in the ball.
12^2 × 3.1416 = 452.39 square inches of surface of the ball.
And, 52.5787 ÷ 452.39 = .1162 + = $\frac{1}{8}$ inch thick, full.   Ans.

To cut a square sheet of copper, tin, etc., so as to form a vessel of the greatest cubical capacity the sheet admits of.

RULE. — From each corner of the sheet, at right angles to the side, cut $\frac{1}{6}$ part of the length of the side, and turn up the sides till the corners meet.

COMPARATIVE COHESIVE FORCE.

Comparative Cohesive Force of Metals, Woods, and other substances, Wrought Iron (medium quality) being the unit of comparison, or 1; the cohesive force of which is 60000 lbs. per inch. transverse area.

Wrought iron,	1.00	Ash, white,	.23
" " wire,	1.71	" red,	.30
Copper, cast,	.40	Beech,	.19
" wire,	.76	Birch,	.25
Gold, cast,	.34	Box,	.33
" wire,	.51	Cedar,	.19
Iron, cast, (average).	.38	Chestnut, sweet,	.17
Lead, "	.015	Cypress,	.10
" milled,	.055	Elm,	.22
Platinum, wire,	.88	Locust,	.34
Silver, cast,	.66	Mahogany, best,	.36
" wire,	.68	Maple,	.18
Steel, soft,	2.00	Oak, Amer., white,	.19
" fine,	2.25	Pine, pitch,	.20
Tin, cast block,	.083	Sycamore,	.22
Zinc, "	.043	Walnut,	.30
" sheet,	.27	Willow,	.22
Brass, cast,	.75	Ivory,	.27
Gun metal,	.50	Whalebone,	.13
Gold 5, copper 1,	.83	Marble,	.15
Silver 5, " 1,	.80	Glass, plate,	.10
Brick,	.05	Hemp fibres, glued,	1.53
Slate,	.20		

The *strength* of white oak to cast iron, is as 2 to 9.
The *stiffness* of   "   "   "   " is as 1 to 13.

To determine the weight, or force, in pounds, necessary to tear asunder a bar, rod, or piece of any of the above named substances, of any given transverse area:

RULE.—Multiply the comparative cohesive force of the substance, as given in the table, by the cohesive force per square inch, area of cross section (60000 lbs.) of wrought iron, which gives the cohesive force of 1 square inch area of cross section of the substance whose power is sought to be ascertained, and the product of 1 square inch thus found, multiplied by area of cross section, in inches, of the rod, piece, or bar itself, gives the cohesive force thereof.

Alloys having a tenarity greater than the sum of their constituents.
Swedish copper 6 pts., Malacca tin 1; tenacity per sq. inch, 64000 lbs.
Chili copper 6 pts., Malacca tin 1;   "   "   " 60000 "
Japan copper 5 pts., Banca tin 1;   "   "   " 57000 "
Anglesea copper 6 pts., Cornish tin 1; "   "   " 41000 "

LINEAR DILATION OF SOLIDS BY HEAT. 75

Common block-tin 4 pts., lead 1, zinc 1; tenacity per sq. in., 13000 lbs.
Malacca tin 4 pts., regulus of antimony 1; "   "   "   12000 "
Block-tin 3 pts., lead 1 part;   "   "   "   10000 "
Block-tin 8 pts., zinc 1 part;   "   "   "   10200 "
Zinc 1 part, lead 1 part;   "   "   "   4500 "

Alloys having a density greater than the mean of their constituents.
GOLD with *antimony, bismuth, cobalt, tin,* or *zinc.*
SILVER with *antimony, bismuth, lead, tin,* or *zinc.*
COPPER with *bismuth, palladium, tin,* or *zinc.*
LEAD with *antimony.*
PLATINUM with *molybdinum.*
PALLADIUM with *bismuth.*

Alloys having a density less than the mean of their constituents.
GOLD with *copper, iron, iridium, lead, nickel,* or *silver.*
SILVER with *copper* or *lead.*
IRON with *antimony, bismuth,* or *lead.*
TIN with *antimony, lead,* or *palladium.*
NICKEL with *arsenic.*
ZINC with *antimony.*

RELATIVE POWER OF DIFFERENT METALS TO CONDUCT ELECTRICITY,

(the mass of each being equal.)

Copper,	 1000	Platinum,	. . .	188
Gold,	 936	Iron,	. . .	158
Silver,	 736	Tin,	. . .	155
Zinc,	 285	Lead,	. . .	83

LINEAR DILATION OF SOLIDS BY HEAT.

Length which a bar heated to 212° *has greater than when at the temperature of* 32°.

Brass, cast,	. . .0018671	Iron, wrought,	. .	.0012575
Copper,	. . .0017674	Lead,	. .	.0028568
Fire brick,	. . .0004928	Marble,	. .	.0011016
Glass,	. . .0008545	Platinum,	. .	.0009342
Gold,	. . .0014880	Silver,	. .	.0020205
Granite,	. . .0007894	Steel,	. .	.0011898
Iron, cast,	. . .0011111	Zinc,	. .	.0029420

NOTE.—To find the *surface* dilation of any particular article, double its linear dilation, and to find the dilation in *volume*, triple it. To find the elongation in linear inches per linear foot, of any particular article, multiply its respective linear dilation, as given in the TABLE, by 12.

MELTING POINT OF METALS AND OTHER BODIES.

Lime, palladium, platinum, porcelain, rhodium, silex, may be melted by means of strong lenses, or by the hydro-oxygen blowpipe. Cobalt, manganese, plaster of Paris, pottery, iron, nickel, &c., at from 2700° to 3250° Fahrenheit; others as follows:—

	Degrees Fah.		Degrees Fah.
Antimony,	809	Nitre,	660
Beeswax, bleached,	155	Silver,	1873
Bismuth,	506	Solder, common,	475
Brass,	1900	" plumbers',	360
Copper,	1996	Sugar,	400
Glass, flint,	1178	Sulphur,	226
Gold,	2016	Tallow,	127
Lead,	612	Tin,	442
Mercury,	—39	Zinc,	680
Cast iron thoroughly melts at			2786
Greatest heat of a smith's forge, (com.)			2346
Welding heat of iron,			1892
Iron red hot in twilight,			884
Lead 1, tin 1, bismuth 4, melts at			201
Lead 2, tin 3, bismuth 5, " "			212

RELATIVE POWER OF DIFFERENT BODIES TO RADIATE HEAT.

Water,	100	Lead, bright,	19
Copper,	12	Mercury,	20
Glass,	90	Paper, white,	100
Ice,	85	Silver,	12
India ink,	88	Tin, blackened,	100
Iron, polished,	15	" clean,	12
Lampblack,	100	" scraped,	16

NOTE.—The power of a body to *reflect* heat is inverse to its power of radiation.

BOILING POINT OF LIQUIDS.
Barometer at 30 in.

Acid, nitric,	253°	Oils, essential, avg.,	318°
" sulphuric,	600°	" turpentine,	316°
Alcohol, *anhyd.*,	168.5°	" linseed,	640°
" 90 per cent.,	174°	Phosphorus,	554°
Ether, sulph.,	97°	Sulphur,	560°
Mercury,	656°	Water,	212°

NOTE.—Barometer at 31 inches, water boils at 213°.57; at 29, it boils at 210°.39; at 28, it boils at 208°.69; at 27, it boils at 206°.65, and in *vacuo* it boils at 88°. No liquid, under pressure of the atmosphere alone, can be heated above its boiling point. At that point the steam emitted sustains the weight of the atmosphere.

FREEZING POINT OF LIQUIDS.

Acid, nitric,	—55°	Oil, linseed, avg., . —11°
" sulphuric,	1°	Proof spirits, . . —7°
Ether,	—47°	Spirits turpentine, . 16°
Mercury,	—39°	Vinegar, . . . 28°
Milk,	30°	Water, . . . 32°
Oil, cinnamon,	30°	Wine, strong, . . 20°
" fennel,	14°	Rapeseed Oil, . . 25°
" olive,	36°	

NOTE. — Water expands in freezing .11, or $\frac{1}{9}$ its bulk.

EXPANSION OF FLUIDS BY BEING HEATED FROM 32° TO 212°, F.

Atmospheric air, $\frac{1}{480}$ per each degree, = .375
Gases, all kinds, $\frac{1}{480}$ "   "   " .
Mercury, exposed, ...018
Muriatic acid, (sp. gr. 1.137,) 060
Nitric acid, (sp. gr. 1.40,) 110
Sulphuric acid, (sp. gr. 1.85,) 060
    " ether, — to its boiling point, . . .070
Alcohol, (90 per cent.,)   "   " . . .110
Oils, fixed, ...080
" turpentine, ...070
Water, . •.. 046

RELATIVE POWER OF SUBSTANCES TO CONDUCT HEAT.

Gold,	1000	Zinc,		363
Silver,	973	Tin,		304
Copper,	898	Lead,		180
Platinum,	381	Porcelain,		12
Iron,	374	Fire brick,		11

NOTE. — Different woods have a conducting power in ratio to each other, as is their respective specific gravities, the more dense having the greater.

METALS IN ORDER OF DUCTILITY AND MALLEABILITY.

Ductility.
1. Platinum.
2. Gold.
3. Silver.
4. Iron.
5. Copper.
6. Zinc.
7. Tin.
8. Lead.

Malleability.
1. Gold.
2. Silver.
3. Copper.
4. Tin.
5. Platinum.
6. Lead.
7. Zinc.
8. Iron.

78 NUTRITIVE AND ALCOHOLIC PROPERTIES OF BODIES.

Quantity per cent. by weight of Nutritious Matter contained in different articles of Food.

Articles.	per ct.	Articles.	per ct.
Lentils,	94	Oats,	74
Peas,	93	Meats, avg.,	35
Beans,	92	Potatoes,	25
Corn, (maize,)	89	Beets,	14
Wheat,	85	Carrots,	10
Barley,	83	Cabbage,	7
Rice,	88	Greens,	6
Rye,	79	Turnips, white,	4

Specific gravity, and quantity per cent., by volume, of Absolute Alcohol contained, necessary to constitute the following named unadulterated articles.

Articles.	Sp. grav. 60°, b. 30.	Per cent. of Alcohol.
Absolute Alcohol, (*anhydrous,*)	.7939	100
Alcohol, highest by distillation,	.825	92.6
" commercial standard,	.8335	90
Proof Spirits, — standard,	.9254	54

Quantity per cent., by volume, (general average) of Absolute Alcohol contained in different pure or unadulterated Liquors, Wines, &c.

Liquors, &c.	per cent.	Wines.	per cent.
Rum,	50	Port,	22
Brandy,	50	Madeira,	20
Gin, Holland,	48	Sherry,	18
Whiskey, Scotch,	50	Lisbon,	17
" Irish,	50	Claret,	10
Cider, whole,	9	Malaga,	16
Ale,	8	Champagne,	14
Porter,	7	Burgundy,	12
Brown Stout,	6	Muscat,	17
Perry,	9	Currant,	19

Proof of Spirituous Liquors.

The weight, in air, of a cubic inch of *Proof Spirits*, at 60° F., is 233 grains; therefore, an inch cube of any heavy body, at that temperature, weighing 233 grains less in spirits than in air, shows the spirits in which it is weighed to be *proof*. If the body lose less of its weight, the spirit is above proof, — if more, it is below.

COMPARATIVE WEIGHT OF TIMBER. 79

Comparative Weight of different kinds of Timber in a green and perfectly seasoned state.

Assuming the weight of each kind destitute of water to be 100, that of the same kind green is as follows:—

Ash,	.	153	Cedar, . .	148	Maple, red,	.	149
Beech,	.	174	Elm, swamp, .	198	Oak, Am.,	.	151
Birch,	.	169	Fir, Amer., .	171	Pine, white,	.	152

NOTE.— Woods which have been felled, cleft and housed for 12 months, still retain from 20 to 25 per cent. of water. They therefore contain but from 75 to 80 per cent. of heating matter; and it will require from 23 to 29 per cent. the weight of such woods to dispel the water they contain. They are, therefore, less valuable by weight, as fuel, by this per cent., than woods perfectly free from moisture. They never, however, contain, exposed to an ordinary atmosphere, less than 10 per cent. of water, however long kept; and even though rendered anhydrous by a strong heat, they again imbibe, on exposure to the atmosphere, from 10 to 12 per cent. of dampness.

Relative power of different seasoned Woods, Coals, &c., as fuel, to produce heat, — the Woods supposed to be seasoned to mean dryness, ($77\frac{1}{2}$ per cent.,) and the other articles to contain but their usual quantity of moisture.

	Ratio of Heating Power per equal	
	Bulk.	Weight.
Hickory, shell-bark, ...	1.00	1.00
" red-heart, ...	.81	.99
Walnut, com.	.95	.98
Beech, red, ...	.74	.99
Chestnut, ...	.49	.98
Elm, white, ...	.58	.98
Maple, hard, ...	.66	.98
Oak, white, ...	.81	.99
" red, ...	.69	.99
Pine, white, ...	.42	1.01
" yellow, ...	.48	1.03
Birch, black, ...	.63	.99
" white, ...	.48	.99
Coal, Cumberland, (bit.)	2.56	2.28
" Lackawanna (anth.)	2.28	2.22
" Lehigh, "	2.39	2.03
" Newcastle, (bit.)	2.10	1.96
" Pictou, (bit.)	2.21	1.91
" Pittsburgh, (bit.)	1.78	1.82
" Peach Mountain, (anth.) . . .	2.69	2.29
Charcoal, ...	1.14	2.53
Coke, Virginia, natural, ...	1.89	2.12
" Cumberland, ...	1.31	2.25
Peat, ordinary, ...		.62
Alcohol, common, ...		2.02
Beeswax, yellow, ...		2.90
Tallow, ...		3.10

NOTE.—By help of the preceding table, the price of either one article being known, the relative or par value of either other, as fuel, may be readily ascertained:—EXAMPLE:
Maple (66) : $5.00 : : Pine (42) : $3.18.

ILLUMINATION — ARTIFICIAL.

The following TABLE shows:—
1. The materials and methods of using — column *Materials*.
2. The comparative maximum intensity of light afforded by each material, used or consumed as indicated, — column *Intensities*.
3. The weight, in grains, of material consumed per hour, by each method respectively, in producing its respective light, or light of intensity ascribed — column *Weight*.
4. The *ratio* of weight required of each material, under each special method of consumption, for the production of equal lights in equal times — column *Ratios*.

Materials.			Intens.	Weight.	Ratios.
Camphene,* Paragon Lamp,			16.	853	1.
Sperm Oil, Parker's heating Lamp,			11.	696	1.19
" " Mech. or Carcel "			10.	815	1.53
" " French annular "			5.	543	2.04
" " Common hand "			1.	112	2.10
Whale " p'f'd., P's heating "			9.	780	1.63
Wax Candles, 3's or 4's, 15 in. or 12 in.,			1.	125	2.35
" " 6's, 9 in.,			.92	122	2.50
Sperm " 4's, 13½ in.,			1.	142	2.66
Stearine " 4's, 13½ in.,			1.	168	3.15
Tallow, " dipped, 10's,			.70	150	4.02
" " mould, 10's,			.66	132	3.75
" " " 8's,			.57	132	4.35
" " " 6's,			.79	163	3.87
" " " 4's, 13¼ in.,			1.	186	3.49
"*Coal Gas*," intensity being			1.	740	

NOTE. — The consumption of 1.43 cubic feet of gas per hour, gives a light equal to one wax candle, — the consumption of 1.96 cubic feet per hour, a light equal to four wax candles, and the consumption of 3 cubic feet per hour, a light equal to ten wax candles. A cubic foot of gas weighs 518 grains.

The average yield of bi-carbureted hydrogen — Olefiant gas — Coal Gas, obtained from the following articles, is as annexed.

    1 lb. Bituminous Coal, ... 4½ cubic feet.
    1 lb. Oil, or Oleine, ... 15 " "
    1 lb. Tar, ... 12 " "
    1 lb. Rosin, or Pitch, ... 10 " "

A pipe whose interior diameter is ⅜ inch, will supply gas equal in illuminating power to 20 wax candles.

    *1 lb. Camphene, . . . = 1 $\frac{1}{10}$ pint.
    1 lb. Sperm Oil, . . . = 1 $\frac{1}{10}$ "
    1 lb. Whale " p'f'd., . . = 1 $\frac{1}{25}$ "

ILLUMINATION. 81

By the foregoing TABLE, it is readily seen in what ratio the several intensities, furnished by the different methods, stand one to another, — that the French annular lamp, for instance, has a maximum power = half that of the mechanical, or $\frac{8}{16}$ that of the camphene, or 5 wax candles, 3 to the lb., — that the camphene, at its maximum power, yields an intensity equal to that afforded by 16 ÷ .57, = 28 tallow candles, moulds, 8 to the lb., — that as the intensity of a six wax candle, 13 in., is .92, and that of an eight mould tallow .57, 57 candles of the former yield an intensity equal to that afforded by 92 of the latter, &c., &c.

THE QUANTITY OF MATERIAL consumed in any given time by either of the foregoing methods, in the production of any given intensity of light, is readily ascertained by help of the preceding TABLE. Suppose, for example, an intensity equal to that afforded by 1 camphene paragon lamp at its greatest power, is required, and for three hours, and that it is proposed to produce the same by tallow candles, moulds, 10 to the lb.; the quantity by weight of candles consumed in the production is required, and, consequently, the number of lights that must be used.

ILLUSTRATION.

Intens. of camph., (16) ÷ intensity of candles, (.66) = 24 candles, and grs. in 1 h. by 1 candle, (132) × 24 × 3 hours = 1 lb. 5¾ oz. *Ans.*

THE ECONOMY OF USE, as between any two materials, under either their respective forms, or methods of consumption, for the production of equal lights in equal times, and therefore for the production of any intensity, is also, by help of the given TABLE, easily learned, the market price of both being known; and, thereby, the *per cent.*, if any difference exist, in favor of the more economical, or less expensive of the two, may be found. To illustrate : —

1. The price of camphene is 10 cents a pound, and that of sperm oil, 15; the economy of use as between the two for the production of equal lights — equal intensities in equal times, greater or less — the former consumed in the paragon lamp, and the latter in Parker's heating, is desired, and the *per cent.* in favor of the less expensive.

10 × 1 = 10, and 15 × 1.19 = 17.85; showing the economy to be in favor of the camphene — showing it so to an extent 17.85 — 10 = $\frac{785}{1785}$, or to an extent 7 cts. 8½ mills per 17 cts. 8½ mills — to an extent, therefore,

17.85 : 7.85 : : 100 = 44 per cent. *Ans.*

2. The price of sperm candles, 4's, is 40 cts. a pound, and the price of tallow, mould 10's, is 11 cts. It is desired to know which of the two, for the production of an intensity nearest obtainable to that afforded by one of the wax, but not less than that of 1 wax, is the less expensive, and to what extent *per cent.*

By casting the eye to the table, it is readily seen that two tallow candles must be employed, the comparative intensity of which is .66 each; therefore, .66 × 2 = 1.32, and 3.75 × 1.32 = 4.95, equivalent weight; consequently—

40 × 2.66 = 106.4, and 11 × 4.95 = 54.45, and 106.4 — 54.45 = 51.95. Therefore,

$$106.4 : 51.95 :: 100 = 48\tfrac{8}{10} \text{ per cent. } Ans.$$

Showing that an intensity nearly ½ greater is afforded by the tallow than the wax, and at an expense 49 *per cent.* less. The same rule of practice is applicable, as between any two methods, for equal or greater or less intensities, as desired.

THERMOMETERS.

	Boiling point.	Freezing point.
Fahrenheit's,	212°	32°
Reaumur's,	80°	0°
Centigrade,	100°	0°

To reduce Reaumur to Fahrenheit.

When it is desired to reduce the +°, (degrees above the zero) :—

RULE.— Multiply the degrees Reaumur, by 2.25, and add 32° to the product; the sum will be the degrees Fahrenheit.

When it is desired to reduce the —°, (degrees below the zero) :—

RULE. — Multiply the —° Reaumur by 2.25, and subtract the product from 32°; the difference will be the degrees Fahrenheit.

EXAMPLE. — The degrees R. are 40; required the equivalent degrees F.

$$40 \times 2.25 = 90 + 32 = 122°. \ Ans.$$

EXAMPLE. — The degrees below 0, R., are 10; what are the corresponding degrees F.?

$$10 \times 2.25 = 22.5, \text{ and } 32 - 22.5 = 9\tfrac{1}{2}°. \ Ans.$$

EXAMPLE. — The degrees below 0, R., are 16; what point on the scale F. corresponds thereto?

$$16 \times 2.25 = 36, \text{ and } 32 - 36 = -4: 4° \text{ below } 0. \ Ans.$$

To reduce the Centigrade to Fahrenheit.

RULE. — Multiply the degrees C. by 1.8, and in all other respects proceed as directed for Reaumur, above.

*NOTE.—The *zero* of Wedgewood's pyrometer is fixed at the temperature of iron red-hot in daylight, = 1077° F., and each degree W. equals 130° F. The instrument is not considered reliable, and is but little used.

HORSE POWER.

A HORSE-POWER, in machinery, as a measure of force, is estimated equal to the raising of 33000 lbs. over a single pulley one foot a minute, = 550 lbs. raised one foot a second, = 1000 lbs. raised 33 feet a minute.

ANIMAL POWER.

A man of ordinary strength is supposed capable of exerting a force of 30 lbs. for 10 hours in a day, at a velocity of 2½ feet a second, = 75 lbs. raised 1 foot a second.

The ordinary working power of a horse is calculated at 750 lbs. for 8 hours in a day, at a velocity of 2 feet a second, = 375 lbs. raised 1 foot a second, = 5 times the effective power of a man during associated labor, and 4 times his power per day; and as machinery may be supposed to work continually, = a trifle less than 23 per cent. per day of a machine horse-power.

STEAM.

Table exhibiting the expansive force and various conditions of steam under different degrees of temperature.

Degrees of heat.	Pressure in atmospheres.	Density. Water as 1.	Volume. Water as 1.	Spec. gravity. Air as 1.	Weight of a cubic foot in grains.
212	1	.00059	1694	.484	254
250.5	2	.00110	909	.915	483
276	3	.00160	625	1.330	700
293.8	4	.00210	476	1.728	910
308	5	.00258	387	2.120	1110
359	10	.00492	203	3.970	2100
418.5	20	.00973	106	7.440	3940

[An atmosphere is $14\frac{7}{10}$ lbs. to the square inch.]

NOTE. — By the above table it is seen that any given quantity of steam having a temperature of 212° F., occupies a space, under the ordinary pressure of the atmosphere, 1694 times greater than it occupied when as water in a natural state. It exerts a mechanical force, consequently, = 1694 times the weight or force of the atmosphere resting on the bulk from which it was generated, or resting on 1-1694th of the space it occupies. A force, if we consider the volume as so many cubic inches, equal to the raising of 2087 lbs. 12 inches high, by a quantity of steam less than a cubic foot, heated only to the temperature of boiling water, and weighing but 248 grains, and that, too, the product of a single cubic inch of water.

The mean pressure of the atmosphere at the earth's surface is equal to the weight of a column of mercury 29.9 inches in height, or to a column of water 33.87 feet in height, = 2116.8 lbs. per square foot, or

14.7 lbs. per square inch. Its density above the earth is uniformly less as its altitude is greater, and its extent is not above 50 miles — its mean altitude is about 45 miles; at 44 miles it ceases to reflect light. Were it of uniform density throughout, and of that at the surface, its altitude would be but 5¼ miles. Its weight is to pure water of equal temperature and volume, as 1 to 829. It revolves with the earth, and its average humidity, at 40° of latitude, is 4 grains per cubic foot. Its weight at 60°, b. 30, compared with an equal bulk of pure water at 40°, b. 30, is as 1 to 830.1.

VELOCITY AND FORCE OF WIND.

Appellations.	Mean velocity in Miles per hour.	Feet per second.	Force in lbs. per square foot.
Just perceptible,	2½	3¾	.032
Gentle, pleasant wind,	4½	6¾	.101
Pleasant, brisk gale,	12½	18¼	.80
Very brisk, "	22½	33	2.52
High wind,	32½	47¾	5.23
Very high wind,	42½	62¼	8.92
Storm, or tempest,	50	73¼	12.30
Great storm,	60	88	17.71
Hurricane,	80	117¼	31.49
Tornado, moving buildings, &c.,	100	146.7	49.20

The curvature of the earth is 6.99 inches (.5825 foot) in a single statute mile, or 8.05 inches in a geographical mile, and is as the square of the distance for any distance greater or less, or space between two levels; thus, for three statute miles it is

$$1 : 3^2 :: 6.99 : 5\tfrac{1}{4} \text{ feet, nearly.}$$

The horizontal refraction is $\tfrac{1}{13}$.

Degrees of longitude are to each other in length, as the cosines of their latitudes. At the equator a degree of longitude is 60 geographical miles in length, at 90° of latitude it is 0; consequently, a degree of longitude at

5°	= 59.77 miles.	40° = 45.96 miles.
10°	= 59.09 "	50° = 38.57 "
20°	= 56.38 "	70° = 20.52 "
30°	= 51.96 "	85° = 5.23 "

Time is to longitude 4 minutes to a degree, — faster, east of any given point; slower, west.

The mean velocity of sound at the temperature of 33° is 1100 feet a second. Its velocity is increased ½ a foot a second for every degree above 33°, and decreased $\frac{1}{4}$ a foot a second for every degree below 33.

In water, sound passes at the rate of 4,708 feet a second.
Light travels at the rate of 192,000 miles per second.

GRAVITATION.

GRAVITY, or GRAVITATION, is a property of all bodies, by which they mutually attract each other proportionally to their masses, and inversely as the square of the distance of their centres apart. Practically, therefore, with reference to our Earth and the bodies upon or near its surface, gravity is a constant force centred at the Earth's centre, and is there continually operating to draw all bodies with a uniformly accelerating velocity to that point, and through very nearly equal spaces, in equal intervals of time from rest, at all localities.

Putting R' to represent the Equatorial radius of the earth, and r to represent the Polar, and making $R' = 3962.5$ statute miles, and $r = 3949.5$, which is nearly in accordance with the mean of the most reliable measurements of the arcs of a degree of latitude at different localities, we have $e^2 = (R'^2 - r^2) \div R'^2 = .006550751$, the square of the ellipticity of the earth, and $R = 2R' \div (2 + e^2 \sin^2 l)$, the radius at any given latitude l.

And since the initial velocity due to gravity at the level of the sea at the Equator is $G = 32.0741$ feet per second, or, in other words, since a body falling in vacuo at the equator, at the level of the sea, describes a space of 16.03705 feet in the first second of time from rest, we have $g = [R' \sqrt{G} \div R]^2$, the initial velocity at the level of the sea at any given radius R; or $g = (22441.2 \div R)^2$.

And finally $g = \left(\dfrac{22441.2}{R}\right)^2 \times \left(1 - \dfrac{2h}{5280R}\right)$ at any given radius R, at any given altitude, h, in feet, above the level of the sea.

NOTE.—When l, reckoned from the equator, is higher than 45°, $\sin^2 l = \cos^2 (90 - l)$.

The *momentum*, or force, with which a falling body strikes, is the product of its weight and velocity (the weight multiplied by the square root of the product of the space fallen through and 64.33, or 4 times $16\frac{1}{12}$); thus, 100 lbs., falling 50 feet, will strike with a force,

50 × 64.333 = $\sqrt{}$ 3216.66 = 56.71 × 100 = 5671 lbs.

An entire revolution of the earth, from west to east, is performed in 23 hours, 56 minutes, and 4 seconds. A solar year = 365 days, 5 hours, 48 minutes, 57 seconds.

The area of the earth is nearly 197,000,000 square miles. Its crust is supposed to be about 30 miles in thickness, and its mean density 5 times that of water. About $\frac{3}{4}$ of its area, or 150,000,000 square miles, is covered by water. The portions of land in the several divisions, in square miles, are, in round numbers, as follows, viz: —

Asia, . . 16,300,000	Europe, . .	3,700,000
Africa, . . 11,000,000	Australia, . .	3,000,000
America, . . 14,500,000		

America is 9000 miles long, or $\frac{35}{100}$ the circumference of the earth.

The population of the globe is about 1,000,000,000, of which there are, in

Asia, . . 456,000,000	Africa, . .	62,000,000
Europe, . . 258,000,000	America, . .	55,000,000

CHEMICAL ELEMENTS.

The chemical elements — simple substances in nature — as far as has been determined, are 58 in number: 13 non-metallic and 45 metallic.

Of the non-metallic, 5 — *bromine, chlorine, fluorine, iodine,* and *oxygen*, (formerly termed "*supporters of combustion*,") have an intense affinity for all the others, which they penetrate, corrode, and apparently consume, always with the production, to some extent, of light and heat. They are all non-conductors of electricity and negative electrics.

The remaining 8 — *hydrogen, nitrogen* or *azote, carbon, boron, silicon, phosphorus, selenium,* and *sulphur,* are eminently susceptible of the impressions of the preceding five; when acted upon by either of them to a certain extent, light and heat are manifestly evolved, and they are thereby converted into incombustible compounds.

Of the metals, 7 — *potassium, sodium, calcium, barytium, lithium, strontium,* and *magnesia,* by the action of oxygen, are converted into bodies possessed of *alkaline* properties.

Seven of them — *glucinum, erbium, terbium, yttrium, allumium, zirconium,* and *thorium,* — by the action of oxygen, are converted into the *earths* proper.

In short, all the metals are acted upon by oxygen, as also by most or all of the non-metallic family. The compounds thus formed are *alkaline, saline,* or *acidulous,* or an *alkali,* a *salt,* or an *acid,* according to the nature of the materials and the extent of combination.

Metals combine with each other, forming *alloys.* If one of the metals in combination is mercury, the compound is called an *amalgam.*

Silicon is the base of the mineral world, and *carbon* of the organized.

For a very general list of the metals, see TABLE OF SPECIFIC GRAVITIES.

TABLE

Exhibiting the Elementary Constituents and per cent. by weight of each, in 100 parts of different compounds.

Compounds.		Hydrogen.	Oxygen.	Azote.	Carbon.
Atmospheric air, . . .	a		20.8	79.2	
Water, pure, . . .		11.1	88.9		
Alcohol, anhydrous, . .		12.9	34.44		52.66
Olive oil, ...		13.4	9.4		77.2
Sperm "		10.97	10.13		78.9
Castor "		10.3	15.7		74.00
Stearine, (solid of fats,) .		11.23	6.3	0.30	82.17
Oleine, (liquid of fats,) .		11.54	12.07	0.35	76.03
Linseed oil, ...		11.35	12.64		76.01
Oil of turpentine, . .		11.74	3.66		84.6
"*Camphene,*" (pure spts. turp.)		11.5			88.5
Caoutchouc, (gum elastic,) .		10.			90.
Camphor, ...		11.14	11.48		77.38
Copal, resin, . . .		9.	11.1		79.9
Guaiac, resin, . . .		7.05	25.07		67.88
Wax, yellow, . . .		11.37	7.94		80.69
Coals, cannel, . . .		3.93	21.05	2.80	72.22
" Cumberland, . .		3.02	14.42	2.56	80.
" Anthracite, . .	b				93.
Charcoal, ...					97.
Diamond, ...					100.
Oak wood, dry, . . .	c	5.69	41.78		52.53
Beech " " . . .		5.82	42.73		51.45
Acetic acid, dry, . .		5.82	46.64		47.54
Citric " crystals, . .		4.5	59.7°		35.8
Oxalic " •dry, . .			79.67		20.33
Malic, " crystals, . .		3.51.	55.02		41.47
Tartaric " dry, . .		3.	60.2		36.80
Formic " " . .		2.68	64.78		32.54
Tannin, tannic acid, solid, .		4.20	44.24		51.56
Nitric acid, dry, . . .			73.85	26.15	
Nitrous " anhydrous, liquid,			61.32	30.68	
Ammoniacal gas, . .		17.47		82.53	
Carbonic acid " . .			72.32		27.68
Carb. hydrogen gas, . .		24.51			75.49
Bi-carb. hyd., olefient gas, .		14.05			85.95
Cyanogen " " .				53.8	46.2
Nitric oxyde " .			53.	47.00	
Nitrous " " .			36.36	63.64	
Ether, sulphuric, . . .		13.85	21.24		65.05
Creosote, ...		7.8	16.		76.2

CONSTITUENTS OF BODIES.

Compounds.		Hydrogen.	Oxygen.	Azote.	Carbon.
Morphia,		6.37	16.29	5.	72.34
Quina, — quinine,		7.52	8.61	8.11	75.76
Veratrine,		8.55	19.61	5.05	66.79
Indigo,		4.38	14.25	10.	71.37
Silk, pure white,		3.94	34.04	11.33	50.69
Starch, — farina, dextrine,		6.8	49.7		43.5
Sugar,		6.29	50.33		43.38
Gluten,		7.8	22.	14.5	55.7
Wheat,	c	6.	44.4	2.4	47.02
Rye,		5.7	45.3	1.7	47.03
Oats,		6.6	38.2	2.3	52.9
Potatoes,		6.1	46.4	1.06	45.9
Peas,		6.4	41.3	4.3	48.
Beet root,		6.2	46.3	1.8	45.7
Turnips,		6.	45.9	1.8	46.3
Fibrin,	d	7.03	20.30	19.31	53.36
Gelatin,	d	7.91	27.21	17.	47.88
Albumen,	d	7.54	23.88	15.70	52.88

Muriatic acid gas, — Hydrogen 5.53 + 94.47 chlorine.
Sulphuric acid, dry, — Oxygen 79.67 + 20.33 sulphur.
Silicic acid — Silica, dry, — Oxygen 51.96 + 48.04 silicon.
Boracic acid — Borax, dry, —   "   68.81 + 31.19 boron.

a. The atmosphere, in addition to its constituents as given in the table, contains, besides a small quantity of vapor, from 1 to 3 parts in a thousand of carbonic acid gas, and a trace merely of ammoniacal gas.

b. Anthracite coal, charcoal, plumbago, coke, &c., have no other constituent than carbon; they are combined, to a small extent, with foreign matters, such as iron, silica, sulphur, alumina, &c.

c. The constituents of woods, grains, &c., are given per cent., without regard to the foreign matters (*metallic*) which they contain. In oak, chestnut, and Norway pine, the ashes amount to about $\frac{4}{10}$ of 1 per cent., and in ash and maple to $\frac{7}{10}$ of 1. In anthracite coals, at an average, they are about 7 per cent.

d. Fibrin, Gelatin, Albumen — Proximate animal constituents — Nutritious properties of animal matter.

Fibrin is the basis of the muscle (lean meat) of all animals, and is also a large constituent of the blood.

Gelatin exists largely in the skin, cartilages, ligaments, tendons and bones of animals. It also exists in the muscles and the membranes.

Albumen exists in the skin, glands and vessels, and in the serum of the blood. It constitutes nearly the whole of the white of an egg.

CONSTITUENTS OF BODIES.   89

The RELATIVE QUANTITIES BY VOLUME of the several gases going to constitute any particular compound, are readily ascertained by help of their respective specific gravities, compared with their relative weights, as given per cent. in the preceding table :—thus, the sp. gr. of hydrogen is .0689, and that of oxygen 1.1025, and 1.1025 ÷ .0689 = 16; showing the weight of the latter to be 16 times that of the former per equal volumes, or, relatively, as 16 to 1. The per cent. by weight, as shown by the table, in which these two gases combine to form water, for instance, is 11.1 and 88.9; or 11.1 of hydrogen and 88.9 of oxygen in 100 of the compound; or as 88.9 ÷ 11.1,—as 8 to 1: 16 ÷ 8 = 2: two volumes, therefore, of the lighter gas (hydrogen) combine with one of oxygen to form water. Water, consequently, is a Protoxide of Hydrogen.

Upon the principle of ATOMIC WEIGHTS — primal quantities, by weight, in which bodies combine, based upon some fixed radix, usually hydrogen as it forms with water, and as 1,— we have, for water,— $H^1 + O^1 =$ Aq. 9. An atom of hydrogen, therefore, is 1, an atom of oxygen 8, and an atom of water 9.

By the same rule as the preceding, the constituents of atmospheric air are found to be to each other in volume as 4 to 1,—4 volumes of nitrogen and 1 volume of oxygen = atmospheric air. The weight of nitrogen to hydrogen per equal volumes, is as 14.14 to 1. Atomically, therefore, oxygen being 8, it is as 7.07 to 1; hence we have $N^4 + O = 36.28$, the atomic weight of atmosphere.

The vast condensation of the gases which takes place, in some instances, in forming compounds, may be conceived of, and the process for determining the same exhibited by a single illustration. We will take, for example, water. A single cubic inch of distilled water, at 60°, weighs 252.48 grains. Its weight is to that of dry atmosphere, at the same temperature, as 827.8 to 1. A cubic inch of dry atmosphere, therefore, at that density, weighs .305 of a grain. Hydrogen, we find by the table of Specific Gravities, weighs .0689 as much as atmosphere, and oxygen 1.1025 as much. A cubic inch of hydrogen, therefore, weighs $.0689 \times .305 = .0210145$ of a grain, and a cubic inch of oxygen $1.1025 \times .305, = .3362625$ of a grain. The constituents of water by volume are 2 of the first mentioned gas to 1 of the latter; and $.0210145 \times 2 + .3362625 = .3782915$ of a grain, = weight of three cubic inches of the uncondensed compound, ⅓ of which, .1260972 of a grain, is the weight of a volume 1 cubic inch.

As the weight of a given volume of the uncondensed compound, is to the weight of an equal volume of the condensed compound, so are their respective volumes, inversely: then —

.1260972 : 252.48 :: 1 : 2002.26, the number of cubic inches of the two gases condensed into 1 inch to form water; a condensation of 2001 times. Of this volume of gases, ⅔, or 1334.84 cubic inches, is hydrogen; the remaining third, 667.42 cubic inches, is oxygen.

The foregoing method, though strictly correct, does not exhibit in a general way the most expeditious for solving questions of that nature, the condensation which takes place in the gases on being converted into solids, or dense compounds. It was resorted to, in part, as a means through which to exhibit principles and proportions pertaining thereto.

As before; one cubic inch of water weighs 252.48 grains, $\frac{1}{9}$ of which, or 28.05+ grains, is hydrogen, and $\frac{8}{9}$, or 224.43— grains, is oxygen. The volume of 1 grain of oxygen is 2.97+ cubic inches, and the volume of hydrogen is 16 times as much, or 47.58+ cubic inches. Therefore, 28.05 × 47.58 = 1334.62, and 224.43 × 2.97 = 665.56, = 2001.18, condensation, as before.

Properties of the SIMPLE SUBSTANCES, *and some of their compounds, not given in the foregoing.*

BROMINE, — at common temperatures, a deep reddish-brown volatile liquid; taste caustic; odor rank; boils at 116°; congeals at 4°; exists in sea-water, in many salt and mineral springs, and in most marine plants; action upon the animal system very energetic and poisonous — a single drop placed upon the beak of a bird destroys the bird almost instantly. A lighted taper, enveloped in its fumes, burns with a flame green at the base and red at the top; powdered tin or antimony brought in contact is instantly inflamed; potash is exploded with violence.

CHLORINE, — a greenish-yellow, dense gas; taste astringent; odor pungent and disagreeable; by a pressure of 60 lbs. to the square inch is reduced to a liquid, and thence, by a reduction of the temperature below 32°, into a solid. It exists largely in sea-water — is a constituent of common salt, and forms compounds with many minerals; is deleterious, irritating to the lungs, and corrosive; has eminent bleaching properties, and is the greatest disinfecting agent known; a lighted taper immersed in it burns with a red flame; pulverized antimony is inflamed on coming in contact, so is linen saturated with oil of turpentine; phosphorus is ignited by it, and burns, while immersed, with a pale-green flame; with hydrogen, mixed measure for measure, it is highly explosive and dangerous.

FLUORINE, — a gas, similar to chlorine, — exists abundantly in *fluor-spar.*

OXYGEN, — a transparent, colorless, tasteless, inodorous, innoxious gas; supports respiration and combustion, but will not sustain life for any length of time, if breathed in a pure state. It is by far the most abundant substance in existence; constitutes $\frac{1}{5}$ of the atmosphere;

⅔ of water; and nearly the whole crust of the earth is oxidized substances. For further combinations and properties, see tables of *Elementary Constituents* and *Chemical Elements*.

IODINE, — at common temperatures, a soft, pliable, opaque, bluish-black solid; taste acrid; odor pungent and unpleasant; fuses at 225°; boils at 347°; its vapor is of a beautiful violet color; it inflames phosphorus, and is an energetic poison; exists mainly in sea-weeds and sponges.

HYDROGEN, — a transparent, colorless, tasteless, inodorous, innoxious gas; if pure, will not support respiration; if mixed with oxygen, produces a profound sleep; exists largely in water; is the basis of most liquids, and is by far the lightest substance known; burns in the atmosphere with a pale, bluish light; mixed with common air, 1 measure to 3, it is explosive; mixed with oxygen, 2 measures to 1, it is violently so.

NITROGEN, or *Azote*, — a transparent, colorless, tasteless, inodorous gas; will not support respiration or combustion, if pure; exists largely as a constituent of the atmosphere — in animals, and in fungous plants; is evolved from some hot springs; in connection with some bodies, appears combustible.

CARBON, — the *diamond* is the only pure carbon in existence; pure carbon cannot be formed by art; *charcoal* is 97 per cent. carbon; *plumbago*, 95; *anthracite*, 93. Carbon is supposed by some to be the *hardest* substance in nature. A piece of charcoal will scratch glass; but it is doubtful if this is not due to the form of its crystals, rather than to the first mentioned quality. It is doubtless the most *durable*. For combinations, &c., see table.

BORON, — a tasteless, inodorous, dark olive-colored solid.

SILICON, — a tasteless, inodorous solid, of a dark-brown color; exists largely in soils, quartz, flint, rock-crystal, &c.; burns readily in air — vividly in oxygen gas; explodes with soda, potassa, barryta.

PHOSPHORUS, — a transparent, nearly colorless solid, of a wax-like texture; fuses at 108°, and at 550° is converted into a vapor; exists mainly in bones — most abundant in those of man — is poisonous; at common temperatures it is luminous in the dark, and by friction is instantly ignited, burning with an intense, hot, white flame; must be kept immersed in water.

SELENIUM, — a tasteless, inodorous, opaque, brittle, lead-colored solid, in the mass; in powder, a deep-red color; becomes fluid at 216°, boils at 650°; vapor, a deep yellow; exists but sparingly, mainly in combination with volcanic matter; is found in small quantities combined with the ores of lead, silver, copper, mercury.

Ammoniacal gas, — $N + H^3$; transparent, colorless, highly pungent and stimulating; alkaline; is converted into a transparent liquid by a pressure of 6.5 atmospheres, at 50°; does not support respiration; is inflammable.

Carbonic acid gas, — $C + O^2$; transparent, colorless, inodorous, dense; is converted into a liquid by a pressure of 36 atmospheres; exists extensively in nature, in mines, deep wells, pits; is evolved from the earth, from ordinary combustion, especially from the combustion of charcoal, and from many mineral springs; is expired by man and animals; forms 44 per cent. of the *carbonate of lime* called marble; the brisk, sparkling appearance of soda-water, and most mineral waters, is due to its presence. It is neither a combustible nor a supporter of combustion; and, when mixed with the atmosphere to an extent in which a candle will not burn, is destructive of life. Being heavier than atmosphere, it may be drawn up from wells in large open buckets; or it may be expelled by exploding gunpowder near the bottom. Large quantities of water thrown in will absorb it.

The above gas is expired by man to the extent of 1632 cubic inches per hour; it is generated by the burning of a wax candle to the extent of 800 cubic inches per hour; and, by the burning of "*Camphene*," (in the production of light equal to that afforded by 1 wax candle,) to the extent of 875 cubic inches per hour. Two burning candles, therefore, vitiate the air to about the same extent as 1 person.

Carbonic oxide gas, — $C + O$; transparent, colorless, insipid; odor offensive; does not support combustion; an animal confined in it soon dies; is highly inflammable, burning with a pale blue flame; mixed with oxygen, 1 to 2, is explosive — with atmosphere, even in small quantity, is productive of giddiness and fainting.

Carburetted hydrogen gas, — $C + H^2$; transparent, colorless, tasteless, nearly inodorous; exists in marshes and stagnant pools — is there formed by the decomposition of vegetable matter; extinguishes all burning bodies, but at the same time is itself highly combustible, burning with a bright but yellowish flame; it is destructive to life, if respired.

Cyanogen — *Bicarburet of Nitrogen* — a gas, — $N + C^2$; transparent, colorless, highly pungent and irritating; under a pressure of

3.6 atmospheres, becomes a limpid liquid; burns with a beautiful purple flame.

Hydrochloric acid gas — *Muriatic acid gas*, — $H + Cl$. (chlorine); transparent, colorless, pungent, acrid, suffocating; strong acid taste.

Nitrous oxide gas — *Protoxide of Nitrogen*, "*laughing gas*," — $N + O$; transparent, colorless, inodorous; taste sweetish; powerful stimulant, when breathed, exciting both to mental and muscular action; can support respiration but from 3 to 4 minutes; is often pernicious in its effects.

Nitric oxide gas — *Binoxide of Nitrogen*, — $N + O^2$; transparent, colorless; wholly irrespirable; lighted charcoal and phosphorus burn in it with increased brilliancy.

Olefiant gas — *Bicarbureted hydrogen gas* — "*coal gas*," — $C^2 + H^2$; transparent, colorless, tasteless, nearly inodorous, when pure; does not support respiration or combustion; a lighted taper immersed in it is immediately extinguished. It burns with a strong, clear, white light; mixed with oxygen, in the proportion of 1 volume to 3, it is highly explosive and dangerous.

Phosphureted hydrogen gas, — $P + H^3$; colorless; odor highly offensive; taste bitter; exists in the vicinity of swamps, marshes, and grave-yards; is formed by the decomposition of bones, mainly; is highly inflammable; takes fire spontaneously on coming in contact with the atmosphere; mixed with pure oxygen, it explodes. It is the veritable "Will o' the wisp."

Sulphureted hydrogen gas — *Hydrosulphuric acid gas*, — $S + H$; transparent, colorless.; taste exceedingly nauseous; odor offensive and disgusting; is furnished by the sulphurets of the metals in general — also by filthy sewers and putrescent eggs. It is very destructive to life; placed on the skin of animals, it proves fatal. It burns with a pale blue flame, and, mixed with pure oxygen, it is explosive.

Hydrocyanic acid — *Prussic acid*, — $N + C^2 + H$; a colorless, limpid, highly volatile liquid; odor strong, but agreeable — similar to that of peach-blossoms; it boils at 79° and congeals at 0; exists in laurel, the bitter almond, peach and peach kernel. It is a most virulent poison, — a drop placed upon a man's arm caused death in a few minutes. A cat, or dog, punctured in the tongue with a needle fresh dipped in it, is almost instantly deprived of life.

Hydrofluoric acid, — $F + H$; a colorless liquid, in well stopped lead or silver bottles, at any temperature between 32° and 59°. It is obtained by the action of sulphuric acid on fluor-spar. It readily acts upon and is used for etching on glass. It is the most destructive to animal matter of any known substance.

Nitrohydrochloric acid —" *aqua regia,*" — (1 part nitric acid and 4 parts muriatic acid, by measure ;) — a solvent for gold. The best solvent for gold is a solution of sal ammoniac in nitric acid.

Nitrosulphuric acid, — (1 part nitric acid and 10 parts sulphuric acid, by measure) — a solvent for silver; scarcely acts upon gold, iron, copper, or lead, unless diluted with water; is used for separating the silver from old plated ware, &c. The best solvent for silver, and one which will not act in the least upon gold, copper, iron, or lead, is a solution of 1 part of nitre in 10 parts of concentrated sulphuric acid, by weight, heated to 160°. This mixture will dissolve about $\frac{1}{4}$ its weight of silver. The silver may be recovered by adding common salt to the solution, and the chloride decomposed by the carbonate of soda.

Selenic acid, — Se $+$ O^3; obtained by fusing nitrate of potassa with selenium — a solvent for gold, iron, copper, and zinc.

Silicic acid, — (*Silica* — silex ; base *Silicon*) — Si $+$ O^3; exists largely in sand. Common glass is fused sand and protoxide of potassium (carbonate of potassa — *potash*) in the proportion of 1 part by weight of the former to 3 of the latter.

Manganese, compounded with oxygen, in different proportions, imparts the various colors and tints given to fancy glass ware, now so generally in vogue.

SECTION III.
PRACTICAL ARITHMETIC.

VULGAR FRACTIONS.

A *fraction* is one or more parts of a UNIT.

A *vulgar fraction* consists of two *terms*, one written above the other, with a line drawn between them.

The term below the line is called the *denominator*, as showing the denomination of the fraction, or number of parts into which the unit is broken.

The term above the line is called the *numerator*, as numbering the parts employed. These together constitute the fraction and its value.

A vulgar fraction always denotes *division*, of which the denominator is the *divisor* and the numerator the *dividend*. Its value as a *unit* is the quotient arising therefrom.

A *simple fraction* is either a proper or improper fraction.

A *proper fraction* is one whose numerator is less than its denominator, as $\frac{1}{2}$, $\frac{2}{3}$, $\frac{24}{50}$, &c.

An *improper fraction* has its numerator equal to or greater than its denominator, as $\frac{2}{2}$, $\frac{4}{3}$, $\frac{24}{4}$, &c.

A *mixed fraction* is a compound of a whole number and a fraction, as $1\frac{1}{2}$, $5\frac{11}{32}$, $12\frac{3}{16}$, &c.

A *compound fraction* is a fraction of a fraction, as $\frac{1}{2}$ of $\frac{3}{4}$; $\frac{1}{2}$ of $\frac{2}{3}$ of $\frac{4}{7}$, &c.

A *complex fraction* has a fraction for its numerator or denominator, or both, as $\frac{\frac{1}{2}}{3}$, $\frac{4}{\frac{2}{3}}$, $\frac{\frac{1}{2}}{\frac{3}{4}}$, $\frac{5\frac{1}{2}}{4}$, &c., and is read: $\frac{1}{2} \div 3$; $4 \div \frac{2}{3}$; $\frac{1}{2} \div \frac{3}{4}$; $5\frac{1}{2} \div 4$, &c.

REDUCTION OF VULGAR FRACTIONS.

To reduce a fraction to its lowest terms.

This consists in concentrating the expression without changing the value of the fraction or the relation of its parts.

It supposes division, and, consequently, by a measure or measures common to both terms.

It is said to be accomplished when no number greater than 1 will divide both terms without a remainder: — therefore,

VULGAR FRACTIONS.

Rule.—Divide both terms by any number that will divide them without a remainder, and the quotient again as before; continue so to do until no number greater than 1 will divide them,—or divide by the greatest common measure at once.

Example.—Reduce $\frac{864}{1512}$ to its lowest terms.

$$4) \frac{864}{1512} = \frac{216}{378} \div 2 = \frac{108}{189} \div 9 = \frac{12}{21} \div 3 = \frac{4}{7}. \text{ Ans.}$$

To reduce an improper fraction to a mixed or whole number.

Rule.—Divide the numerator by the denominator and to the whole number in the quotient annex the remainder, if any, in form of a fraction, making the divisor the denominator as before; then reduce the fraction to its lowest terms.

Example. $\frac{5}{4} = 1\frac{1}{4}$; $\frac{16}{12} = 1\frac{4}{12} = 1\frac{1}{3}$; $\frac{26}{13} = 2$.

To reduce a mixed fraction to an equivalent improper fraction.

Rule.—Multiply the whole number by the denominator of the fractional part, and to the product add the numerator, and place their sum over the said denominator.

Example.—Reduce $3\frac{1}{4}$ and $12\frac{8}{9}$ to improper fractions.

$$3 \times 4 = 12 + 1 = \frac{13}{4}. \text{ Ans.} \qquad 12 \times 9 + 8 = \frac{116}{9}. \text{ Ans.}$$

To reduce a whole number to an equivalent fraction having a given denominator.

Rule.—Multiply the whole number by the given denominator, and place the said denominator under the product.

Example.—How may 8 be converted into a fraction whose denominator is 12?

$$8 \times 12 = \frac{96}{12}. \text{ Ans.}$$

To reduce a compound fraction to a simple one.

Rule.—Multiply all the numerators together for a numerator, and all the denominators together for a denominator; the fraction thus formed will be an equivalent, but often not in its lowest terms. Or, concentrate the expression, when practicable, by reciprocally expunging, or writing out, such factors as exist or are attainable common to both terms, and then multiply the remaining terms as directed above.

Note.—This last practice is called cancellation, or cancelling the terms. It consists, as has been stated, in reciprocally annulling, or casting out, equal values from both terms, whereby the expression is concentrated, and the relation of the parts kept undisturbed; and it may always be carried to the extent of reducing the fraction to its lowest terms, before any multiplication, as final, is resorted to; and often, therefore, to the extent that such multiplication is inadmissible, the terms having been cancelled by each other until but a single number is left in each.

VULGAR FRACTIONS.

EXAMPLE. — Reduce $\frac{2}{3}$ of $\frac{3}{4}$ of $\frac{1}{2}$ to a simple fraction.

Operation by multiplication, $\frac{2}{3} \times \frac{3}{4} \times \frac{1}{2} = \frac{6}{24} = \frac{1}{4}$. *Ans.*

Operation by cancellation, $\dfrac{\cancel{2}\,\cancel{3}\,1}{\cancel{3}\,\cancel{4}\,\cancel{2}} = \frac{1}{4}$. *Ans.*

EXAMPLE. — Reduce $\frac{2}{3}$ of $\frac{3}{4}$ of $1\frac{2}{8}$ of $\frac{6}{8}$ of $\frac{5}{8}$ of 2 to a simple fraction.

By multiplication, $\frac{2}{3} \times \frac{3}{4} \times 1\frac{2}{8} \times \frac{6}{8} \times \frac{5}{8} \times \frac{2}{1} = \frac{4320}{6912} = \frac{5}{8}$. *Ans.*

The last example stated for cancellation, $\left\{ \dfrac{2\ 3\ 12\ 6\ 5\ 2}{3\ 4\ 8\ 8\ 9} \right.$

PROCESS OF CANCELLING THE ABOVE.

1. The 3 in num. equals the 3 in denom., therefore erase both.
2. The first 2 in num. equals or measures the 4 in denom. *twice*, therefore place a 2 under the 4, and erase the 4 and 2 which measured it — (as 4 : 2 : : 2 : 1.)
3. The 2 (remaining factor of 4 and 2 erased) in denom., and the remaining 2 in num., will cancel each other, — erase them.
4. The 12 and 6 in num. = 72, and the 9 and 8 in denom. = 72; these, therefore, in their relations as factors equal each other, and may be erased.

The remaining factors represent the true value of the compound fraction, and will be found = $\frac{5}{8}$, as by multiplication.

EXAMPLE. — Reduce $1\frac{8}{13}$ of $\frac{7}{12}$ to a simple fraction.

$\dfrac{\overset{3}{\cancel{\overset{9}{\cancel{18}}}} \times 7}{13 \times \underset{\underset{2}{\cancel{6}}}{\cancel{12}}}$. Or, $\dfrac{\overset{3}{\cancel{18}} \times 7}{13 \times \underset{2}{\cancel{12}}}$ (= 18 ÷ 6, and 12 ÷ 6) = $\frac{3}{2} \times \frac{7}{13}$ = $\frac{21}{26}$. *Ans.*

To reduce fractions of different denominations to an equivalent simple one, — to a fraction having a common denominator.

RULE. — Multiply each numerator by all the denominators except its own and add the products together for the numerator, and multiply all the denominators together for a denominator.

NOTE. — Whole numbers and fractions other than simple, must first be reduced to simple fractions before they can be reduced to a fraction having a common denominator.

EXAMPLE. — Reduce $\frac{2}{3}$ and $\frac{3}{4}$ to an equivalent simple fraction.

$\frac{2}{3} \times \frac{3}{4} = \frac{8 + 9}{12} = \frac{17}{12}$. *Ans.*

EXAMPLE. — Reduce $\frac{1}{2}$, $\frac{3}{5}$, $\frac{7}{8}$, and $\frac{14}{3}$ to an equivalent.

$\frac{1}{2} + \frac{3}{5} = \frac{11}{10} + \frac{7}{8} = \frac{158}{80} + \frac{14}{3} = \frac{1594}{240} = 6\frac{77}{120}$. *Ans.*

9

VULGAR FRACTIONS.

To reduce a complex fraction to a simple one.

RULE. — Multiply the numerator of the upper fraction by the denominator of the lower, for the new numerator; and the denominator of the upper by the numerator of the lower for the new denominator.

EXAMPLES. — Reduce $\frac{\frac{1}{2}}{3}$, $\frac{4}{\frac{2}{5}}$, $\frac{\frac{1}{2}}{\frac{3}{4}}$, and $\frac{5\frac{1}{2}}{4}$, each to a simple fraction.

$\frac{1}{2} \div \frac{3}{1} = \frac{1}{6}$; $\frac{4}{1} \div \frac{2}{5} = \frac{20}{2}$; $\frac{1}{2} \div \frac{3}{4} = \frac{1}{2} \times \frac{4}{3}, = \frac{4}{6}, = \frac{2}{3}$; $5\frac{1}{2} = \frac{11}{2}$, and $\frac{11}{2} \times \frac{1}{4} = \frac{11}{8}, = 1\frac{3}{8}$. *Ans.*

To reduce Vulgar Fractions to equivalent Decimals.

RULE. — Divide the numerator by the denominator; the quotient is the decimal, or the whole number and decimal, as the case may be.

EXAMPLE. — Reduce $\frac{7}{8}$, $4\frac{3}{5}$, $\frac{14}{12}$, to decimals.

$7 \div 8 = 0.875$; $4\frac{3}{5} = \frac{23}{5}, = 4.6$; $14 \div 12 = 1.166 +$. *Ans.*

To find the greatest common measure or divisor of both terms of a simple fraction, or of two numbers.

RULE. — Divide the greater number by the less; then divide the divisor by the remainder; and so on, continuing to divide the last divisor by the last remainder until nothing remains; the last divisor is the greatest common measure of the two terms.

EXAMPLE. — What is the greatest common measure of $\frac{132}{256}$ or of 132 and 256?

```
132 ) 256 ( 1
      132
      ---
      124 ) 132 ( 1
            124
            ---
              8 ) 124 ( 15
                  120
                  ---
                    4 ) 8 ( 2
                        8
                        -
```
4. *Ans.*

To find the least common denominator of two or more fractions of different denominators, or the least common multiple of two or more numbers.

RULE. — Divide the given denominators, or numbers, by any number greater than 1, that will divide at least two of them without a remainder, which quotient together with the undivided numbers set in a line beneath. Divide the second line as before, and so on, until

VULGAR FRACTIONS.   99

there are no two numbers in the line that can be thus divided; the product of all the divisors and remaining numbers in the last (undivided) dividend is the least common denominator, or multiple sought.

EXAMPLE. — What is the least common denominator of $\frac{7}{20}$, $\frac{4}{25}$, and $\frac{3}{50}$, or of 20, 25, and 50?

$$\begin{array}{r}5\ )\ 20.25.50 \\ \hline 2\ )\ \ 4.5.10 \\ \hline 5\ )\ \ 2.5.5 \\ \hline 2.1.1\end{array}$$

$5 \times 2 \times 5 \times 2 = 100$. *Ans.*

ADDITION OF VULGAR FRACTIONS.

Sum of the products of each numerator with all the denominators except that of the numerator involved, forms numerator of sum.
Product of all the denominators forms denominator of sum.

RULE. — Arrange the several fractions to be added, one after another, in a line from left to right; then multiply the numerator of the first by the denominator of the second, and the denominator of the first by the numerator of the second, and add the two products together for the numerator of the sum; then multiply the two denominators together for its denominator; bring down the next fraction, and proceed in like manner as before, continuing so to do until all the fractions have been brought down and added. Or, reduce all to a common denominator, then add the numerators together for the numerator of the sum, and write the common denominator beneath.

EXAMPLES. — Add together $\frac{1}{2}$, $\frac{2}{3}$, $\frac{3}{4}$, and $\frac{4}{5}$.

$\frac{1}{2} \times \frac{2}{3} = \frac{7}{6} \times \frac{3}{4} = \frac{46}{24} \times \frac{4}{5} = \frac{234}{72} = \frac{13}{4} = 3\frac{1}{4}$. *Ans.*

$\frac{1}{2} = \frac{2}{4} + \frac{3}{4} = \frac{5}{4} = 1\frac{2}{4}$, and $\frac{2}{3} + \frac{4}{5} = \frac{6}{5} = \frac{22}{15}$, and $\frac{12}{15} + \frac{22}{15} = \frac{13}{12}$
$= \frac{13}{4} = 3\frac{1}{4}$. *Ans.*

SUBTRACTION OF VULGAR FRACTIONS.

Product of numerator of minuend and denominator of subtrahend, forms numerator of minuend, for common denominator.
Product of numerator of subtrahend and denominator of minuend, forms numerator of subtrahend, for common denominator.
Product of denominators forms common denominator.
Difference of new found numerators forms the numerator, and common denominator the denominator, of the difference, or remainder sought.

RULE. — Write the subtrahend to the right of the minuend, with the sign (—) between them; then multiply the numerator of the minuend by the denominator of the subtrahend, and the denominator of the minuend by the numerator of the subtrahend; subtract the latter product from the former, and to the remainder or difference affix the product of the two denominators for a denominator; the sum thus formed is the answer, or true difference.

EXAMPLES. — Subtract $\frac{1}{2}$ from $\frac{3}{4}$, also $\frac{2}{5}$ from $\frac{11}{17}$.

$\frac{3}{4} - \frac{1}{2} = \frac{6-4}{8}, = \frac{2}{8} = \frac{1}{4}$. *Ans.*

$\frac{11}{17} - \frac{2}{5} = \frac{55-51}{85} = \frac{4}{85}$. *Ans.*

DIVISION OF VULGAR FRACTIONS.

Product of numerators of dividend and denominators of divisor, forms numerator of quotient.
Product of denominators of dividend and numerators of divisor, forms denominator of quotient; therefore,

RULE. — Write the divisor to the right of the dividend with the sign ($\div$) between them; then multiply the numerator of the dividend by the denominator of the divisor, for the numerator of the quotient, and the denominator of the dividend by the numerator of the divisor, for the denominator of the quotient. Or, invert the divisor, and multiply as in multiplication of fractions. Or, proceed by cancellation, when practicable.

EXAMPLES. — Divide $\frac{1}{2}$ by $\frac{3}{4}$; $\frac{3}{4}$ by $\frac{1}{2}$; $\frac{4}{5}$ by $\frac{11}{13}$; and $\frac{1}{2}$ of $\frac{3}{4}$ of $\frac{5}{6}$ of $\frac{4}{5}$ by $\frac{1}{4}$ of $\frac{1}{2}$ of $\frac{3}{4}$ of $\frac{2}{3}$.

$\frac{1}{2} \div \frac{3}{4} = \frac{4}{6}$; $\frac{3}{4} \div \frac{1}{2} = \frac{6}{4}$; $\frac{4}{5} \div \frac{11}{13} = \frac{52}{55}$; or $\frac{4}{5} \times \frac{13}{11} = \frac{52}{55}$. *Ans.*
$\frac{1}{2} \times \frac{3}{4} \times \frac{5}{6} \times \frac{4}{5} = \frac{60}{144} = \frac{5}{12}$, and $\frac{1}{4} \times \frac{1}{2} \times \frac{3}{4} \times \frac{2}{3} = \frac{6}{96} = \frac{1}{16}$, and $\frac{5}{12} \div \frac{1}{16} = \frac{80}{12} = \frac{20}{3} = 6\frac{2}{3}$. *Ans.*

FORM FOR CANCELLATION. — EXAMPLE LAST GIVEN.

$$\frac{1 \quad 3 \quad 5 \quad 4 \quad\quad 4 \quad 2 \quad 4 \quad 3}{2 \quad 4 \quad 6 \quad 3 \quad\quad 1 \quad 1 \quad 3 \quad 2} = \frac{20}{3}. \text{ Ans., as above.}$$

NOTE. — The foregoing example can be cancelled to the extent of leaving but a 4 and a 5 (= 20) numerators, and a 3 denominator. Units, or 1's, in the expressions, are valueless, as a sum multiplied by 1 is not increased.

MULTIPLICATION OF VULGAR FRACTIONS.

Product of numerators of multiplier and multiplicand, forms numerator of product.
Product of denominators of multiplier and multiplicand, forms denominator of product.

RULE. — Multiply the numerators together for a numerator, and the denominators together for the denominator.

EXAMPLES. — Multiply $\frac{1}{2}$ by $\frac{1}{2}$; $\frac{3}{4}$ by 7; $\frac{11}{12}$ by $\frac{12}{7}$; $\frac{1}{2}$ of $\frac{2}{3}$ of $\frac{3}{4}$ by $\frac{3}{4}$ of $\frac{1}{2}$ of $\frac{2}{3}$.

$\frac{1}{2} \times \frac{1}{2} = \frac{1}{4}$; $\frac{3}{4} \times \frac{7}{1} = \frac{21}{4}$; $\frac{11}{12} \times \frac{12}{7} = \frac{132}{84} = \frac{11}{7}$; $\frac{1}{2} \times \frac{2}{3} \times \frac{3}{4} = \frac{6}{24} = \frac{1}{4}$, and $\frac{3}{4} \times \frac{1}{2} \times \frac{2}{3} = \frac{6}{24} = \frac{1}{4}$, and $\frac{1}{4} \times \frac{1}{4} = \frac{1}{16}$ *Ans*

MULTIPLICATION AND DIVISION OF FRACTIONS, COMBINED.

It has been seen that a compound fraction is converted into an equivalent simple one, by multiplying the numerators together for a numerator, and the denominators together for a denominator; and it has also been seen that a series of simple fractions are converted into a product, by the same process. It is therefore evident that compound fractions and simple, or a series of compound and a series of simple, may be multiplied into each other, for a product, by multiplying all the numerators of both together for a numerator, and all the denominators of both together for a denominator; and that the product will be the same as would be obtained, if the compound were first converted into an equivalent simple fraction, and the simple fractions into a product or factor, and these multiplied together for a product.

It has also been seen that a fraction is divided by a fraction by multiplying the numerator of the dividend by the denominator of the divisor, for the numerator of the quotient, and the denominator of the dividend by the numerator of the divisor, for the denominator of the quotient; and that this multiplication becomes direct as in multiplying for a product, if the divisor is inverted. And it is clear that a compound divisor, or a series of simple divisors, or both, may be used instead of their simple equivalent, and with the same result, if all are inverted.

It is therefore evident that any proposition, or problem, in fractions, consisting of multiplications and divisions both, and these only, no matter how extensive and numerous, or whether in compound fractions, or simple, or both, may be solved, and the true result obtained, as a product, by simply multiplying all the numerators in the statement together for a numerator, and all the denominators in the statement for a denominator, all the divisors in the statement being inverted; that is, all the numerators of the divisors being made denominators in the statement, and all the denominators of the divisor being made numerators in the statement. And it is further evident that a proposition stated in this way, admits of easy cancellation as far as cancellation is practicable, which is often to great extent.

EXAMPLE. — It is required to divide 12 by $\frac{2}{3}$ of $\frac{1}{4}$; to multiply the quotient by the product of 4 and 8; to divide that product by $\frac{7}{2}$ of $\frac{3}{5}$ of 8; to multiply the quotient by $\frac{7}{8}$ of $\frac{8}{9}$ of $\frac{9}{14}$; and to divide that product by the product of 5 and 9.

VULGAR FRACTIONS.

STATEMENT.

(Dividends read from right to left, divisors from left to right.)

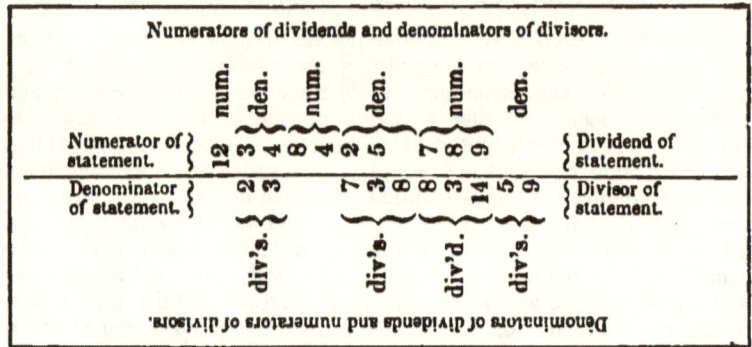

The *answer* to the above proposition is $1\frac{1}{2}\frac{1}{4}$, and the proposition as stated may be readily cancelled to its lowest terms. It may be cancelled to the extent of leaving but 4, 4, 2 in the numerator, and 7, 3, in the denominator, $\frac{4 \times 4 \times 2}{7 \times 3} = \frac{32}{21} = 1\frac{1}{2}\frac{1}{4}$.

To reduce a fraction in a higher denomination to an equivalent fraction in a given lower denomination.

RULE. — Multiply the fraction to be reduced — numerators into numerator and denominators into denominator — by a fraction whose numerator represents the number of parts of the lower denomination, required to make ONE of the denomination to be reduced.

EXAMPLE. — Reduce $\frac{7}{8}$ of a foot to an equivalent fraction in inches.
$\frac{7}{8} \times \frac{12}{1} = \frac{84}{8} = \frac{21}{2}$. *Ans.*

EXAMPLE. — Reduce $\frac{5}{6}$ of a pound to an equivalent fraction in $\frac{2}{3}$ ounces.
$\frac{5}{6} \times \frac{16}{1} = \frac{80}{6} \div \frac{2}{3} = \frac{240}{12} = \frac{20}{1}$. *Ans.*
Or, $\frac{5}{6} \times \frac{16}{1} \times \frac{3}{2} = \frac{240}{12} = \frac{20}{1}$. *Ans.*

To reduce a fraction in a lower denomination to an equivalent fraction in a given higher denomination.

RULE. — Multiply the fraction to be reduced — numerator into denominator and denominator into numerator — by a fraction whose numerator represents the number of parts required of the lower denomination to make 1 of the higher.

EXAMPLE. — Reduce $\frac{21}{2}$ inches to an equivalent fraction in feet.
$\frac{21}{2} \div \frac{12}{1} = \frac{21}{24} = \frac{7}{8}$. *Ans.* Or, $\frac{21}{2} \times \frac{1}{12} = \frac{21}{24} = \frac{7}{8}$. *Ans.*

VULGAR FRACTIONS. 103

EXAMPLE. — Reduce $\frac{40}{2}$ two third ounces to an equivalent fraction in pounds.
$$\frac{40}{2} \times \frac{2}{3} = \frac{80}{6} \div \frac{16}{1} = \frac{80}{96} = \frac{5}{6}. \text{ Ans.}$$
Or, $\frac{40}{2} \times \frac{2}{3} \times \frac{1}{16} = \frac{80}{96} = \frac{5}{6}$. Ans.

To reduce a fraction in a higher to whole numbers in lower denominations.

RULE. — Multiply the numerator of the given fraction by the number of parts of the next lower denomination that make ONE of the given fraction, and divide the product by the denominator. Multiply the numerator of the fractional part of the quotient thus obtained by the number of parts in the next lower denomination that make 1 of the denomination of the quotient, and divide by its denominator for whole numbers as before; so proceed until the whole numbers in each denomination desired are obtained.

EXAMPLE. — How many hours, minutes, and seconds, in $\frac{9}{14}$ of a day?
$$\frac{9}{14} \times 24 = \frac{216}{14} = 15, \tfrac{3}{7} \times 60 = \frac{180}{7} = 25, \tfrac{5}{7} \times 60 = \frac{300}{7} = 42\tfrac{6}{7}, =$$
15 h., 25 m., 42$\tfrac{6}{7}$ sec. Ans.

EXAMPLE. — How many minutes in $\frac{9}{14}$ of a day?
$$\tfrac{9}{14} \times 24 \times 60 = \frac{12960}{14} = 925\tfrac{5}{7}. \text{ Ans.}$$

To reduce fractions, or whole numbers and fractions, in lower denominations, to their value in a higher denomination.

RULE. — Reduce the mixed numbers to improper fractions, find their common denominator, and change each whole number and numerator to correspond therewith. Then reduce the higher numbers to their values in the lowest denomination, add the value in the lowest denomination thereto, and take their sum for a numerator. Multiply the common denominator by the number required of the lowest denomination to make ONE of the next higher, that product by the number required of that denomination to make 1 of the next higher, and so on, until the highest denomination desired is reached, and take the product for a denominator, and reduce to lowest terms.

EXAMPLE. — Reduce 5$\tfrac{1}{3}$ oz., 3$\tfrac{1}{5}$ dwts., 2$\tfrac{1}{2}$ grs., troy, to lbs.
$$\tfrac{16}{3} \cdot \tfrac{16}{5} \cdot \tfrac{5}{2} = \frac{160 \cdot 96 \cdot 75}{30} ; \text{ therefore,}$$

$$160 \times 20 = 3200$$
$$\underline{96}$$
$$3296 \times 24 = 79104$$
$$\underline{75}$$
$$\left. \begin{array}{r} 79179 \\ 30 \times 24 \times 20 \times 12 = \overline{172800} \end{array} \right\} = .458 + \text{ lbs. Ans.}$$

EXAMPLE. — Reduce 11 hours, 59 minutes, 60 seconds, to the fraction of a day.

$$11 \times 60 = 660$$
$$59$$
$$\overline{}$$
$$719 \times 60 = 43140$$
$$60$$
$$\overline{}$$
$$60 \times 60 \times 24 = \left.\frac{43200}{86400}\right\} = \tfrac{1}{2}. \ Ans.$$

EXAMPLE. — Reduce 15 h., 25 m., 42⅚ sec., to the fraction of a day.

$$15 \times 60 \times 60 = 54000$$
$$25 \times 60 = 1500$$
$$42\tfrac{5}{7}$$
$$\overline{}$$
$$55542\tfrac{5}{7}$$
$$7$$
$$\overline{}$$
$$7 \times 60 \times 60 \times 24 = \left.\frac{388800}{604800}\right\} = \tfrac{9}{14}. \ Ans.$$

To work fractions, or whole numbers and fractions, by the Rule of Three, or Proportion.

RULE. — Reduce the mixed terms to simple fractions, state the question as in whole numbers, invert the divisor, and multiply and divide as in whole numbers.

EXAMPLE. — If 2½ yards of cassimere cost $4¼, what will ¾ of a yard cost? $2\tfrac{1}{2} = \tfrac{5}{2}$; $4\tfrac{1}{4} = \tfrac{17}{4}$; then,

$$\tfrac{5}{2} : \tfrac{17}{4} :: \tfrac{3}{4} : x, = \tfrac{17 \times 3 \times 2}{4 \times 4 \times 5} = \tfrac{102}{80} = \$1.27,5. \ Ans.$$

DECIMAL FRACTIONS.

A decimal fraction is written with its numerator only. Its denominator is understood. It occupies one or more places of figures, and has a point or dot (.) prefixed or placed before it. The dot (.) alone distinguishes it from an integer or whole number. It supposes a denominator whose value is a UNIT broken into parts, having a tenfold relation to the number of places the numerator occupies. The denominator, therefore, of any decimal is always a unit (1) with as many ciphers annexed as the numerator has places of figures. Thus, the denominator of .1, .2, .3, &c., is 10, and the fractions are read, *one tenth, two tenths, three tenths*, &c. The denominator of .01, .11, .12, &c., is 100, and these are read, *one hundredth, eleven hundredths,*

DECIMAL FRACTIONS.   105

twelve hundredths, &c. The denominator of .001, .101, .125, &c., is 1000, and these are read *one thousandth, one hundred and one thousandths, one hundred and twenty-five thousandths*, &c. The denominator of a decimal occupying four places of figures as .7525 is 10000, and so on continually.

The first figure on the right of the decimal point is in the place of *tenths*, the second in the place of *tenths* of *tenths*, or *hundredths*, the third in the place of *tenths* of *tenths* of *tenths*, or *thousandths*, &c. Thus the value of a decimal occupying four places of figures, as

.7525, for example, is $\frac{7525}{10000} = \frac{752\frac{1}{2}}{1000} = \frac{75\frac{1}{4}}{100} = \frac{7\frac{1}{2}}{10} + \frac{\frac{1}{4}}{100} = \frac{\frac{3}{4}}{1} + \frac{\frac{1}{4}}{100}$. A decimal is converted into a vulgar fraction of equal value, by affixing its denominator.

Ciphers placed on the right of decimals do not change their value. Thus, .1850 = .185, plainly for the reason that the denominator of the latter bears the same relation to that of the former that 185 bears to 1850; from both terms of the fraction a ten fold has been dropped.

Ciphers placed on the left of decimals *decrease* their value ten fold for every cipher so placed. Thus, .1 = $\frac{1}{10}$, .01 = $\frac{1}{100}$, .001 = $\frac{1}{1000}$, &c.

A *mixed number* is a whole number and a decimal. Thus, 4.25 is a mixed number. Its value is 4 units, or *ones*, and $\frac{25}{100}$ of 1, = $\frac{425}{100}$ = $4\frac{1}{4}$. The number on the left of the separatrix is always a whole number — that on its right, always a decimal.

ADDITION OF DECIMALS.

RULE. — Set the numbers directly under each other according to their values, whole numbers under whole numbers, and decimals under decimals; add as in whole numbers, and point off as many places for decimals in the sum as there are figures in that decimal occupying the greatest number of places.

EXAMPLES. — Add together .125, .34, .1, .8672. Also, 125, 34.11, .235. 1.4322.

```
    .125              125.
    .34                34.11
    .1                  .235
    .8672              1.4322
   ──────             ────────
   1.4322  Ans.       160.7772  Ans.
```

SUBTRACTION OF DECIMALS.

RULE. — Set the numbers, the less under the greater, and in other respects as directed for addition; subtract as in whole numbers, and point off as many places for decimals in the remainder as the decimal having the greatest number of figures occupies places.

EXAMPLES. — Subtract .2653 from .8. Also, 11.5 from 238.134.

```
   .8              238.134
 .2653              11.5
 ─────            ───────
 .5347  Ans.      226.634  Ans.
```

MULTIPLICATION OF DECIMALS.

RULE. — Multiply as in whole numbers, and point off as many places for decimals in the product as there are decimal places in the multiplicand and multiplier both. If the product has not so many places, prefix ciphers to supply the deficiency.

EXAMPLES. — Multiply 14.125 by 3.4. Also, 5.14 by .007.

```
 14.125             5.14
    3.4             .007
 ──────           ──────
 56500            .03598  Ans.
 42375
 ──────
 48.0250 = 48.025.  Ans.
```

NOTE. — Multiplying by a decimal is equivalent to dividing by a whole number that bears the same relation to a UNIT that a unit bears to a decimal. Multiplying by a decimal, therefore, is equivalent to dividing by the denominator of a fraction of equal value whose numerator is 1, or of dividing by the denominator of a fraction of equal value whose numerator is more than 1, and multiplying the quotient by the numerator. Thus, the decimal $.25 = \frac{25}{100} = \frac{1}{4}$, and the decimal $.875 = \frac{875}{1000} = \frac{7}{8}$. And $14.23 \times .25 = 3.5575$, and $14.23 \div 4 = 3.5575$. So, also, $14.23 \times .875 = 12.45125$, and $14.23 \div 8 = 1.77875 \times 7 = 12.43125$. It is sometimes a saving of labor and matter of convenience to achieve multiplication by this process.

DIVISION OF DECIMALS.

RULE. — Write the numbers as for division of whole numbers, then remove the separatrix in the dividend as many places of figures to the right, (supplying the places with ciphers if they are not occupied,) as there are decimal figures in the divisor; consider the divisor a whole number and divide as in division of whole numbers.

EXAMPLE. — Divide .5 by .17. Also, .129 by 4.

```
 .17).50(2.94+.  Ans.     4).129(.032+.  Ans.
      34                      12
     ───                     ───
     160                       9
     153                       8
     ───                     ───
      70                       1
      68
     ───
       2
```

DECIMAL FRACTIONS. 107

Examples. — Divide 16.5 by 1.232.   Also, 1.2145 by 12.231.

```
1.232,)16.500,(13.3928+.   Ans. | 12.231,)1.214,50(.09929+ ) Ans.
       1232                     |          1 100 79  .0993— )
       ----                     |          -------
       4180                     |          113 710
       3696                     |          110 079
       ----                     |          -------
       4840                     |          3 6310
       3696                     |          2 4462
       -----                    |          -------
       11440                    |          1 18480
       11088                    |          1 10079
       -----                    |          -------
       3520                     |          8401
       2464
       -----
       10560
       ,9856
       -----
       704
```

Note. — Dividing by a decimal is equivalent to multiplying by a whole number that bears the same proportion to a UNIT that a unit bears to the decimal. Dividing by a decimal, therefore, is equivalent to multiplying by the denominator of a fraction of equal value whose numerator is 1, or multiplying by the denominator of a fraction of equal value whose numerator is more than 1, and dividing the product by the numerator. Dividing by a fraction is equivalent to multiplying by its denominator and dividing the product by its numerator, or dividing by its numerator and multiplying the quotient by its denominator. Thus, $.5 = \frac{5}{10} = \frac{1}{2}$, and $.75 = \frac{75}{100} = \frac{3}{4}$. And $12.24 \div .5 = 24.48$, and $12.24 \times 2 = 24.48$. So, also, $12.24 \div .75 = 16.32$, and $12.24 \times 4 = 48.96 \div 3 = 16.32$. This method of accomplishing division may often be resorted to with convenience.

REDUCTION OF DECIMALS.

To reduce a decimal in a higher to whole numbers in successive lower denominations.

Rule. — Multiply the decimal by that number in the next lower denomination that equals ONE of the denomination of the decimal, and point off as many places for a remainder as the decimal so multiplied has places. Multiply the remainder by the number in the next lower denomination that equals 1 of the denomination of the remainder, and point off as before; so continue, until the reduction is carried to the lowest denomination required.

Example. — What is the value of .62525 of a dollar?

```
        .      .62525
               100
        Cents, 62.52500
               10
        Mills,  5.25000    Ans. 62 cents 5¼ mills.
```

EXAMPLE. — What is the value of .46325 of a barrel?

$$
\begin{array}{rl}
& .46325 \\
& 32 \\ \hline
\text{Gallons,} & 14.82400 \\
& 4 \\ \hline
\text{Quarts,} & 3.296 \\
& 2 \\ \hline
\text{Pints,} & .592 \\
& 4 \\ \hline
\text{Gills,} & 2.368. \quad Ans. \; 14 \text{ gals. } 3 \text{ qts. } 2\tfrac{368}{1000} \text{ gills.}
\end{array}
$$

EXAMPLE. — How many pence in .875 of a pound?

.875 × 240 = 210. *Ans.*

To reduce decimals, or whole numbers and decimals, in lower denominations, to their value in a higher denomination.

RULE. — Reduce all the given denominations to their value in the lowest denomination, then divide their sum by the number required of the lowest denomination to make ONE of the denomination to which the whole is to be reduced.

EXAMPLE. — Reduce 14 gallons, 3 quarts, 2.368 gills, to the decimal of a barrel.

14 × 4 = 56 + 3 = 59 × 8 = 472 + 2.368 = 474.368.

8 × 4 × 32 = 1024) 474.368 (.46325. *Ans.*

To work decimals, or whole numbers and decimals, by the Rule of Three, or Proportion.

RULE. — State the question and work it as in whole numbers, taking care to point off as many places for decimals in the product to be used as the dividend, as there are decimals in the two terms which form it, and to remove the decimal point therein as many places to the right as there are decimals in the term to be used as a divisor, before the division is had.

EXAMPLE. — If .75 of a pound of copper is worth .31 of a dollar how much is 3.75 lbs. worth?

$$
\begin{array}{r}
.75 : .31 :: 3.75 \\
.31 \\ \hline
375 \\
1125 \\ \hline
.75 \,)\, 1.16,25 \; (\$1.55. \quad Ans.
\end{array}
$$

PROPORTION, OR RULE OF THREE.

The Rule of Proportion involves the employment of three terms — a divisor and two factors for forming a dividend — and seeks a quotient, which, when the proposition is written in ratio, bears the same relation to the third term that the second term bears to the first. Two of the terms given are of like name or nature, and the other is of the name or nature of the quotient or answer sought. That of the nature of the answer is always one of the factors for forming the dividend, and, if the answer is to be greater than that term, the larger of the remaining two is the other; but if the answer is to be less than that term, the less of the remaining two is the other — the remaining term is the divisor.

Example. — If $12 buy 4 yards of cloth, how many yards will $108 buy?

$$\frac{\cancel{4} \times 108}{\cancel{12}_3} = \frac{108}{3} = 36 \text{ yards.} \quad Ans.$$

Example. — If 4 yards of cloth cost $12, how many dollars will 36 yards cost?

$$\frac{12 \times 36}{4} = 108 \text{ dollars.} \quad Ans.$$

Example. — If 30 men can finish a piece of work in 12 days, how many men will be required to finish it in 8 days?

$$\frac{30 \times 12}{8} = 45 \text{ men.} \quad Ans.$$

Example. — If 45 men require 8 days to finish a piece of work, how many men will finish the same work in 12 days?

$$\frac{45 \times 8}{12} = 30 \text{ men.} \quad Ans.$$

Example. — If 8 days are required by 45 men to finish a piece of work, how many days will be required by 30 men to finish the same work?

$$\frac{8 \times 45}{30} = 12 \text{ days.} \quad Ans.$$

Example. — If 12 days are required by 30 men to perform a piece of work, how many days will be required by 45 men to do the same work?

$$\frac{12 \times 30}{45} = 8 \text{ days.} \quad Ans.$$

Example. — I borrowed of my friend $150, which I kept 3 months, and, on returning it, lent him $200; how long may he keep the sum that the interest, at the same rate per cent., may amount to that which his own would have drawn?

$$150 \times 3 \div 200 = 2\tfrac{1}{4} \text{ months. } Ans.$$

EXAMPLE. — A garrison of 250 men is provided with provisions for 30 days, how many men must be sent out that the provisions may last those remaining 42 days?

$$250 \times 30 \div 42 = 179, \text{ and } 250 - 179 = 71. \ Ans.$$

EXAMPLE. — If to the short arm of a lever 2 inches from the fulcrum there be suspended a weight of 100 lbs., what power on the long arm of the lever 20 inches from the fulcrum will be required to raise it?

$$20 : 2 :: 100 = 10 \text{ lbs. } Ans.$$

EXAMPLE. — At what distance from the fulcrum on the long arm of a lever must I place a pound weight, to equipoise or weigh 20 lbs., suspended 2 inches from the fulcrum at the other end?

$$1 : 2 :: 20 : 40 \text{ inches. } Ans.$$

NOTE. — If we examine the foregoing with reference to the fact, we shall see that every proposition in simple proportion consists of a *term and a half!* or, in other words, of a *compound* term consisting of two factors, and a factor for which another factor is sought that together shall equal the compound. We have only to multiply the factors of the compound together — and a little observation will enable us to distinguish it — and divide by the remaining factor, and the work is accomplished. See COMPOUND PROPORTION.

COMPOUND PROPORTION, OR DOUBLE RULE OF THREE.

COMPOUND PROPORTION, like *single* proportion, consists of THREE terms given by which to find a fourth — a divisor and two factors for forming a dividend — but unlike single proportion, one or more of the terms is a compound, or consists of two or more factors; and sometimes a portion of the fourth term is given, which, however, is always a part of the divisor.

Of the given terms, two are suppositive, dissimilar in their natures, and relate to each other, and to each other only; and upon their relation the whole is made to depend; the remaining term is of the nature of one of the former, and relates to the fourth term, which is of the nature of the other.

The object sought is a number, which, multiplied into the factor or factors of the fourth term given, if any, and if not, which of itself, bears the same proportion to the dissimilar term to which it relates, as the suppositive term of like nature bears to the term to which it relates.

RULE. — Observe the denomination in which the demand is made, and of the suppositive terms make that of like nature the second, and the other the first; make the remaining term the third term; and, if

COMPOUND PROPORTION. 111

there are any factors pertaining to the fourth term, affix them to the first; multiply the second and third terms together and divide by the first, and the quotient is the answer, term, or portion of a term, sought.

EXAMPLE.—If 12 horses in 6 days consume 36 bushels of oats, how many bushels will suffice 21 horses 7 days?

$$12 \times 6 : 36 :: 21 \times 7 : x.$$

$$\frac{\overset{3}{\cancel{36}} \times 21 \times 7}{\underset{2}{\cancel{12} \times \cancel{6}}} = \frac{147}{2} = 73\tfrac{1}{2} \text{ bushels. } Ans.$$

EXAMPLE.—If 12 horses in 6 days consume 36 bushels of oats, how many horses will consume 73½ bushels in 7 days?

$$36 : 12 \times 6 :: 73\tfrac{1}{2} : 7 \times x.$$

$$\frac{12 \times 6 \times 73\tfrac{1}{2}}{36 \times 7} = \frac{147}{7} = 21 \text{ horses. } Ans.$$

EXAMPLE.—If the interest on $1 is 1.4 cts. for 73 days, (exact interest at 7 per cent.,) what will be the interest on $150.42 for 146 days?

$$73 : 1.4 :: 150.42 \times 146 : x.$$

$$\frac{1.4 \times 150.42 \times 146}{73} = \$4.21. \ Ans.$$

EXAMPLE.—If the interest on $1 is 1.2 cts. for 73 days, (exact interest at 6 per cent.,) what will be the interest on $125 for 90 days?

$$73 : 1.2 :: 125 \times 90 : x = \$1.85. \ Ans.$$

EXAMPLE.—If $100 at 7 per cent. gain $1.75 in 3 months, how much at 6 per cent. will $170 gain in 11½ months?

$$100 \times 7 \times 3 : 1.75 :: 170 \times 6 \times 11.5 : x.$$

$$1.75 \times 170 \times 6 \times 11.5 \div 100 \times 7 \times 3 = \$9.77,5. \ Ans.$$

EXAMPLE.— By working 10 hours a day 6 men laid 22 rods of wall in 3 days; how many men at that rate, who work but 9 hours a day, will lay 40 rods of wall in 8 days?

$$22 : 6 \times 3 \times 10 :: 40 : 9 \times 8 \times x.$$

$$6 \times 3 \times 10 \times 40 \div 22 \times 9 \times 8 = 4\tfrac{6}{11}. \ Ans.$$

EXAMPLE.—If it costs $112 to keep 16 horses 30 days, and it costs as much to keep 2 horses as it costs to keep 5 oxen, how much will it cost to keep 28 oxen 36 days?

$$16 \times 30 : 112 :: \tfrac{2}{5} \times 28 \times 26 : x.$$
$$\text{Or,} - 16 \times 30 \times 5 : 112 :: 28 \times 36 \times 2 : x.$$

$$\frac{\cancel{112}\ 28\ \cancel{36}^{12}\ \cancel{2}}{\cancel{16}\ \cancel{30}\ 5}^{7} = \frac{28 \times 12 \times 7}{5 \times 5} = \$94.08. \ \textit{Ans.}$$
$$\cancel{15}$$
$$5$$

EXAMPLE. — If 24 men, in 8 days of 10 hours each, can dig a trench 250 feet long, 8 feet wide, and 4 feet deep, how many men, in 12 days of eight hours each, will be required to dig a trench 80 feet long, 6 feet wide, and 4 feet deep?
$$250 \times 8 \times 4 : 24 \times 8 \times 10 :: 80 \times 6 \times 4 : 12 \times 8 \times x = 5-. \ \textit{Ans.}$$

EXAMPLE. — If 120 men in six months perform a given task, working 10 hours a day, how many men will be required to accomplish a like task in 5 months, working 9 hours a day?
$$120 \times 6 \times 10 = 5 \times 9 \times x.$$
$$\text{Or,} - 1 : 120 \times 6 \times 10 :: 1 : 5 \times 9 \times x. = 160. \ \textit{Ans.}$$

EXAMPLE. — The weight of a bar of wrought iron, 1 foot in length, 1 inch in breadth, and 1 inch thick, being 3.38 lbs., (and it is so,) what will be the weight of that bar whose length is $12\tfrac{1}{2}$ feet, breadth $3\tfrac{1}{4}$ inches, and thickness $\tfrac{3}{4}$ of an inch?
$$1 : 3.38 :: 12.5 \times 3.25 \times .75 : x.$$
$$\text{Or,} - 1 : 3.38 :: \tfrac{25}{2} \times \tfrac{13}{4} \times \tfrac{3}{4} : x, \text{ and}$$
$$\frac{3.38 \times 25 \times 13 \times 3}{2 \times 4 \times 4} = 102.98+ \text{ lbs.} \ \textit{Ans.}$$

EXAMPLE. — The weight of a bar of wrought iron, one foot in length and 1 inch square, being 3.38 lbs., what length shall I cut from a bar whose breadth is $2\tfrac{3}{4}$ inches, and thickness $\tfrac{1}{2}$ inch, in order to obtain 10 lbs.? $\quad 3.38 : 1 :: 10 : \tfrac{11}{4} \times \tfrac{1}{2} \times x.$
$$\frac{1 \times 10 \times 4 \times 2}{3.38 \times 11 \times 1} = 2 \text{ feet } 1\tfrac{8}{10} \text{ inches.} \ \textit{Ans.}$$

CONJOINED PROPORTION, OR CHAIN RULE.

THE CHAIN RULE is a process for determining the value of a given quantity in one denomination of value, in some other given denomination of value; or the immediate relationship which exists between two denominations of value, by means of a *chain* of approximate steps,

CONJOINED PROPORTION, OR CHAIN RULE. 113

circumstances, or equivalent values, known to exist, which connect them. In every instance at least *five* terms or values are employed in the process, and in all instances the number employed will be uneven. A proposition involving but three terms, of this nature, is a question in single proportion. The equivalent values employed are divided into *antecedents* and *consequents*, or causes and effects; and the value or quantity for which an equivalent is sought, is called the odd term.

RULE. — 1. *When the value in the denomination of the first antecedent is sought of a given quantity in the denomination of the last consequent.* — Multiply all the antecedents and the odd term together for a dividend, and all the consequents together for a divisor; the quotient will be the answer or equivalent sought.

RULE. — 2. *When the value in the denomination of the last consequent is sought of a given quantity in the denomination of the first antecedent.* — Multiply all the consequents and the odd term together for a dividend, and all the antecedents together for a divisor; the quotient will be the answer required.

EXAMPLE. — I am required to give the value, in Federal money, of 5 Canada shillings, and know no immediate connection or relationship between the two currencies — that of Canada and that of the United States. The nearest that I do know is that 20 Canada shillings have a value equal to 32 New York shillings, and that 12 New York shillings equal in value 9 New England shillings, and that 15 New England shillings equal $2.50; and with this knowledge will seek the value, in Federal money, of the 5 Canada shillings.

$$\frac{2.50 \times 9 \times 32 \times 5}{15 \times 12 \times 20} = \$1. \text{ Ans.}$$

EXAMPLE. — If $2½ equal 15 New England shillings, and nine shillings in New England equal 12 shillings in New York, and 32 shillings in New York equal 20 shillings in Canada, how many shillings in Canada will equal $1?

$$\frac{15 \quad \overset{3}{\cancel{12}} \quad \overset{8}{\cancel{20}} \quad 1}{\underset{3}{\cancel{2½}} \quad \underset{}{\cancel{9}} \quad \underset{4}{\cancel{32}}} = \tfrac{15}{3} = 5 \text{ shillings. Ans.}$$

EXAMPLE. — If 14 bushels of wheat weigh as much as 15 bushels of fine salt, and 10 bushels of fine salt as much as 7 bushels of coarse, and 7 bushels of coarse salt as much as 4 bushels of sand, how many bushels of sand will weigh as much as 40 bushels of wheat?

$$\frac{15 \times 7 \times 4 \times 40}{14 \times 10 \times 7} = 17\tfrac{1}{7} \text{ bushels. Ans.}$$

PERCENTAGE.

Pure percentage, or PERCENTAGE, is a rate by the hundred of a *part* of a quantity or number denominated the principal, or basis. But percentage, considered as a means, and as commonly applied, is mixed and related in an eminent degree; and in this light may be regarded as divided into orders bearing different names.

Thus *Interest* is percentage related to intervals of time in the past.

Discount is percentage related to interest, and intervals of time in the future.

Profit and Loss is comparative percentage, or percentage related to the positive and negative interests in business, etc., etc.

Pure percentage is commonly called BROKERAGE when paid to a broker for services in his line.

It is called COMMISSION when paid to or received by a factor or commission merchant for buying or selling goods.

It is called PREMIUM by an insurance company, when taken for insuring against loss.

It is called PRIMAGE when it is a charge in addition to the freight of a vessel, etc.

Comparative percentage relates to the differences of quantities, and is confined always to the idea of *more* or *less*. It implies ratio. This description of percentage, though much in practice, seems not to be well understood; and often a quantity is indirectly stated to be many times less than nothing, or many times greater than it is. The difference of two quantities cannot be as great as a hundred per cent. of the greater, however widely unequal the quantities may be, nor as small as no per cent. of the greater or lesser, however nearly equal they may be. No quantity or number can be as small as 1 time less than another quantity or number; and therefore cannot be as small as 100 per cent. less. But, since one quantity may be many by 1 time, or many times greater than another with which it is compared, it may be said to be many by 100 times, or many hundred per cent. greater.

When one of two quantities in comparison is stated to be three times less, or three hundred per cent. less, for instance, than the other, the expression is incorrect and absurd. The meaning evidently is, that it is two-thirds less, or only one-third as large as the other, — that it is $66\frac{2}{3}$ per cent. less, or only $33\frac{1}{3}$ per cent. as large as the other. In common comparison, 1 is the measuring unit. In percentage, 100 is the measuring unit.

PERCENTAGE.

Let $a =$ principal.
    $b =$ percentage.
    $s =$ amount (sum of the principal and percentage).
    $d =$ difference of the principal and percentage.
    $r =$ rate of the percentage.
    $p =$ rate per cent. of the percentage.

$a = s - b = b \div r = 100b \div p = 100s \div (100 + p),$
$b = s - a = ar = ap \div 100,$
$p = 100r = 100b \div a = 100(s - a) \div a,$
$r = p \div 100 = b \div a = (s - a) \div a,$
$s = a + b = a(1 + r) = a(100 + p) \div 100,$
$d = a - b = 2a - s = s - 2b = a(1 - r).$

To find the Percentage.

EXAMPLES.

What is $\frac{1}{4}$ of 1 per cent. of $200?
$$b = ar = ap \div 100 = \$0.50. \quad Ans.$$

$\frac{8}{7}$ of 2 per cent. of 50 is what part of 50?
$$\frac{50 \times 8 \times 2}{7 \times 100} = 1\frac{1}{7}. \quad Ans.$$

What is $\frac{2}{3}$ of $\frac{2}{3}$ of $\frac{1}{2}$ of 24 per cent. of 150 lbs.?
$$150 \times 12 \div 100 = 18 \text{ lbs.} \quad Ans.$$

What is $2\frac{3}{8}$ per cent. of 19 bushels?
$$\tfrac{19}{8} \times \tfrac{19}{100} = 0.45125 \text{ bushels.} \quad Ans.$$

Bought a job lot of merchandise for $850, and sold it the same day, brokerage, $2\frac{1}{2}$ per cent., for $975; what was the net gain?
$$s - sr - a = s - (sr + a) = s(1 - r) - a = 975 - 975 \times .025 - 850 = \$100.625. \quad Ans.$$

To find the Rate or Rate Per Cent.

EXAMPLES.

What per cent. of $20 is $2?
$$r = b \div a, \; p = 100b \div a = 10 \text{ per cent.} \quad Ans.$$

12 dozen is equal to what per cent. of 2 dozen?
$$12 \div 2 = 6, \; 600 \text{ per cent.} \quad Ans.$$

PERCENTAGE.

What part of $5\frac{1}{4}$ lbs. is $\frac{3}{4}$ of 2 lbs.?

$$r = \frac{3}{4} \times \frac{4}{2\mathbf{1}} = \frac{1\cdot 4}{4\cdot 2\mathbf{1}} = 0.27\frac{3}{2\mathbf{1}}. \quad Ans.$$

$24\frac{1}{2}$ per cent. is what per cent. of $36\frac{3}{4}$ per cent.?

$$66\frac{2}{3} \text{ per cent.} \quad Ans.$$

For an article that cost $4, $5 were received; what per cent. of $4 was received?

$$p = 5 \times 100 \div 4 = 125 \text{ per cent.} \quad Ans.$$

A farmer sowed 4 bushels of wheat, which produced 48 bushels; what per cent. was the *increase*? 48 is *more* than 4 by what per cent. of 4? The difference of 48 and 4 is what per cent. of 4?

$$r = \frac{a-b}{b} = \frac{a}{b} - 1, \ p = \frac{100(a-b)}{p} = \frac{48-4}{4} = 48 \div 4 - 1 =$$

$$100(48-4) \div 4 = 1100 \text{ per cent.} \quad Ans.$$

What per cent. would have been the *decrease*, if he had sowed 48 bushels, and harvested only 4 bushels? 4 is *less* than 48 by what rate of 48? The difference of 48 and 4 is what per cent. of 48?

$$r = (a-b) \div a = 1 - \frac{b}{a} = 0.91\frac{2}{3}, \text{ or } 91\frac{2}{3} \text{ per cent.} \quad Ans.$$

Since water is composed of 8 atoms of oxygen and 1 atom of hydrogen, what per cent. of it is oxygen? 8 is what per cent. of the sum of 8 and 1?

$$r = \frac{a}{a+b} = 1 - \frac{b}{a+b}, \ p = \frac{100a}{a+b} = \frac{8}{8+1} = .8889-,$$

$$\text{or } 88.89 - \text{per cent.} \quad Ans.$$

What per cent. of it is hydrogen? 1 is what per cent. of the sum of 8 and 1?

$$r = 1 - \frac{a}{a+b} = \frac{b}{a+b}, \ p = \frac{100b}{a+b} = \frac{1}{8+1} = .1111 +, \text{ or}$$

$$11.11 + \text{ per cent.} \quad Ans.$$

How many volumes of water must be added to 100 volumes of 90 per cent. alcohol to reduce it to 50 per cent. alcohol or common proof? 90 is more than 50 by what per cent. of 50? The difference of 90 and 50 is what per cent. of 50?

$$p = \frac{(a-b)100}{b} = \frac{(90-50)100}{50} = 80. \quad Ans.$$

PERCENTAGE.   117

How many volumes of 50 per cent. alcohol must be added to 100 volumes of 90 per cent. alcohol to produce 80 per cent. alcohol? 90 is more than 80 by what per cent. of the difference of 80 and 50? The difference of 90 and 80 is what per cent. of the difference of 80 and 50?

$$p = \frac{(a-b)100}{b-b'} = \frac{(90-80)100}{80-50} = 33\frac{1}{3}. \ Ans.$$

How many volumes of 90 per cent. alcohol must be added to 100 volumes of 50 per cent. alcohol to raise it to 80 per cent. alcohol? 50 is less than 80 by what per cent. of the difference of 90 and 80? The difference of 80 and 50 is what per cent. of the difference of 90 and 80?

$$\frac{(b-b')100}{a-b} = \frac{(80-50)100}{90-80} = 300. \ Ans.$$

If to 2 volumes of 95 per cent. alcohol, 1 volume of 50 per cent. alcohol be added, what per cent. alcohol will be the mixture? The sum of 50 and twice 95 is what per cent. of the sum of 2 and 1?

$$\frac{2a+b}{2+1} = \frac{2 \times 95 + 50}{2+1} = 80 \text{ per cent.} \ Ans.$$

In a barrel of apples, the number of sound ones was 60 per cent. *greater* than the number that were damaged. What per cent. *less* was the number that were damaged than the number that were sound? 60 per cent. is what per cent. of the sum of 100 per cent. and 60 per cent.? .6 is what rate of $1 + .6$?

$$r = \frac{a}{1+a} = 1 - \frac{100}{1+a} = \frac{0.a}{1+.a} = 1 - \frac{1}{1.a} = \frac{60}{1+60} = .375, \text{ or}$$

$37\frac{1}{2}$ per cent. *Ans.*

Since the number of damaged apples was $37\frac{1}{2}$ per cent. less than the number that were sound, what per cent. greater was the number that were sound than the number that were damaged?

$$r = a \div (1-a) = 1 \div (1-a) - 1 = 60 \text{ per cent.} \ Ans.$$

Since the number of sound ones was 60 per cent. greater than the number that were damaged, what per cent. of the whole were sound?

$$r = \frac{a+a^2}{2a} = \frac{1+.a}{2}, \ p = \frac{100+60}{2} = 80 \text{ per cent.} \ Ans.$$

What per cent. of the whole were damaged?

$$(100-60) \div 2 = 20 \text{ per cent.} \ Ans.$$

118 PERCENTAGE.

Since 20 per cent. of the apples were damaged, what per cent. less was the number that were damaged than the number that were sound?

$$r = \frac{1 - 2.a}{2 - 2.a} = 1 - \frac{1}{2 - 2.a}, \; p = \frac{100 - 2a}{200 - 2a} = 100 - \frac{100}{200 - 40} = 37\tfrac{1}{2} \text{ per cent. } Ans.$$

What per cent. greater was the number that were sound than the number that were damaged?

$$r = 2 - (1 + 2.a) = 2 - 2.a - 1 = 60 \text{ per cent. } Ans.$$

Since 80 per cent. of the whole were sound, what per cent. less was the number that were damaged than the number that were sound?

$$r = \frac{2.a - 1}{2.a} = 1 - \frac{1}{2.a} = \frac{2 \times .80 - 1}{2 \times .80} = 37\tfrac{1}{2} \text{ per cent. } Ans.$$

Since the number of damaged ones was $37\tfrac{1}{2}$ per cent. less than the number that were sound, what per cent. of the whole were sound?

$$r = \frac{1}{2 - 2:a}, \; p = \frac{100}{2 - 2a} = \frac{100}{2 - 2 \times 37.5} = 80 \text{ per cent. } Ans.$$

Since 80 per cent. of the whole were sound, what per cent. greater was the number that were sound than the number that were damaged?

$$r = \frac{2 - .a}{2} = 2.a - 1 = 2 \times .80 - 1 = 60 \text{ per cent. } Ans.$$

Lost 20 per cent. of a cargo of coal by jettison, and 5 per cent. of the remainder by screening, what per cent. of the coal was saved?

$$\left. \begin{array}{l} a - b' = d' \\ d' - b'' = d'' \\ d'' - b''' = d''', \text{ \&c.} \end{array} \right\} \; r = (1 - r')(1 - r'') = (1 - .20) - (1 - .20) \times .05 = (1 - .20)(1 - .05) = 76 \text{ per cent. } Ans.$$

Yesterday drew 12 per cent. of my balance of $4,273 in the bank, and deposited $1,000; and to-day have drawn $31\tfrac{1}{4}$ per cent. of the balance left over, or as it stood last night. What per cent. of the sum of the first-mentioned balance and deposit of yesterday have I drawn?

$$r = \frac{b' + b''}{a + m} = \frac{512 + 1487.575}{4273 + 1000} = 37.9354 + \text{ per cent. } Ans.$$

PERCENTAGE. 119

What per cent. of the said sum is remaining in the bank ?

$$1 - \frac{b' + b''}{a + m} = \frac{a + m - b' - b''}{a + m} = \frac{a + m - (b' + b'')}{a + m} = 62.0646 - \text{ per cent. } Ans.$$

What per cent., predicating it upon the first-mentioned balance, have I drawn ?

$$r = \frac{b' + b''}{a} = \frac{512.76 + 1487.576}{4273} = 46.8134 - \text{ per cent. } Ans.$$

What per cent. have I drawn, predicating it upon what I now have in the bank ?

$$r = \frac{b' + b''}{a - b' + m - b''} = \frac{b' + b''}{a + m - (b' + b'')} = 61.1225 + \text{ per cent. } Ans.$$

What amount of money must I deposit to make good $62\frac{1}{4}$ per cent. of the aforementioned sum ?

$$d = r(a + m) + b' + b'' - (a + m) = r(a + m) - d'' = \$22.96. \ Ans.$$

To find the Principal or Basis.

EXAMPLES.

The percentage being 250, and the rate .06, what is the principal ?

$$a = b \div r = 100b \div p = 250 \div .06 = 25,000 \div 6 = 4,166\tfrac{2}{3}. \ Ans.$$

A tax at the rate of $\frac{5}{8}$ of 1 per cent. on the valuation was $27.50. What was the valuation ?

$$a = \frac{b \times 6 \times 100}{5} = \$3,300. \ Ans.$$

Sold 120 barrels of flour, which amounted to 12 per cent. of a certain consignment. The consignment consisted of how many barrels ?

$$120 \div 0.12 = 1,000. \ Ans.$$

216 bushels is *more* by 8 per cent., or 8 per cent. more, than what number of bushels ? 8 per cent. more than what number is equal to 216 ? What number, plus 8 per cent. of it, will make 216 ?

$$a = s \div (1 + r) = 216 \div 1.08 = 200. \ Ans.$$

200 lbs. is *less* by 8 per cent., or 8 per cent. less, than what num-

ber of lbs.? 8 per cent. less than what number is 200? What number, minus 8 per cent. of it, is equal to 200?

$$a = d \div (1-r) = 200 \div (1-.08) = 217\tfrac{9}{23}. \quad Ans.$$

$$\therefore 217\tfrac{9}{23} - 217\tfrac{9}{23} \times .08 = 200 = a - b = d = a(1-r).$$

To a quantity of silver, a quantity of copper equal to 20 per cent. of the silver is to be added, and the mass is to weigh 22 ounces. What weight of silver is required?

$$a = s \div (1+r) = 22 \div 1.2 = 18\tfrac{1}{3} \text{ ounces.} \quad Ans.$$

What weight of copper is required?

$$s - \frac{s}{1+r} = \frac{sr}{1+r} = 3\tfrac{2}{3} \text{ ounces.} \quad Ans.$$

To a quantity of copper, a quantity of nickel equal to $62\tfrac{1}{2}$ per cent. of the copper, a quantity of zinc equal to $33\tfrac{1}{3}$ per cent. of the copper, and a quantity of lead equal to 5 per cent. of the copper, are to be added; and the whole is to weigh $40\tfrac{1}{6}$ pounds. The weight of each constituent of the alloy is required.

$$a = \frac{s}{1+r+r'+r''} = \frac{40\tfrac{1}{6}}{1+.62\tfrac{1}{2}+.33\tfrac{1}{3}+.05}$$

$$\left. \begin{array}{l} = 20 \text{ lbs. of copper,} \\ b = 20\,r = 12\tfrac{1}{2} \text{ lbs. of nickel,} \\ b' = 20\,r' = 6\tfrac{2}{3} \text{ lbs. of zinc,} \\ b'' = 20\,r'' = 1 \text{ lb. of lead.} \end{array} \right\} Ans.$$

INTEREST.

Universal for any rate per cent.

$T =$ time in months and decimal parts of a month; $t =$ time in days; $P =$ principal; $r =$ rate per cent., expressed decimally; $i =$ interest.

$$i = \frac{P \times T \times r}{12} = \frac{P \times t \times r}{365}.$$

$$P = \frac{12\,i}{Tr} = \frac{365\,i}{tr}. \quad T = \frac{12\,i}{Pr}. \quad t = \frac{365\,i}{Pr}. \quad r = \frac{12\,i}{PT} = \frac{365\,i}{Pt}.$$

EXAMPLE. — A promissory note, made April 27, 1864, for

INTEREST.   121

825\frac{25}{100}$ and interest at 6 per cent., matured Oct. 6, 1865 : what was the interest?

<table>
<tr><td colspan="4">Oct. is 10th month.
April is 4th month.</td><td>Time from April 27 to Oct. 6 (one of the dates always included) = 162 days, which, added to the 365 days in the year preceding = 527 days.</td></tr>
<tr><td></td><td>r.</td><td>m.</td><td>d.</td><td></td></tr>
<tr><td></td><td>1865</td><td>. 10 .</td><td>6</td><td></td></tr>
<tr><td></td><td>'64</td><td>. 4 .</td><td>27</td><td></td></tr>
<tr><td>Time =</td><td>1</td><td>. 5 .</td><td>9</td><td>NOTE.—One day's interest at least is generally lost by computing the time in years and months, or months, instead of days.</td></tr>
</table>

$$825.25 \times 17.3 \times .06 \div 12 = \$71.38. \quad Ans.$$

$$825.25 \times 527 \times .06 \div 365 = \$71.49. \quad Ans.$$

To find a constant divisor, k, for any given rate per cent.

When the time is taken in months, $k = 12 \div r$.
When the time is taken in days, $k = 365 \div r$; thus,

When the RATE is 6 per cent. $\frac{P \times t}{6083}$ = Interest.

When the RATE is 7 per cent. $\frac{P \times t}{5214}$ = Interest, &c.

EXAMPLE.— Required the interest on $750 for 93 days, at 7 per cent.

$$750 \times 93 \div 5214 = \$13.38. \quad Ans.$$

EXAMPLE.— What is the rate per cent. when $450 gains 94\frac{1}{2}$ in 3 years?

$$450 : 100 :: 94.5 : 3x = 7 \text{ per cent.} \quad Ans.$$

$$94.5 \div 3 \times 450 = .07. \quad Ans.$$

EXAMPLE.— In what time will $125 at 6 per cent. gain 18\frac{3}{4}$?

$$6 : 100 :: 18.75 : 125 \times x = 2\tfrac{1}{2} \text{ years.} \quad Ans.$$

$$18.75 \div 125 \times .06 = 2\tfrac{1}{2} \text{ years.} \quad Ans.$$

EXAMPLE.— What principal at 5 per cent. interest will gain 16\frac{7}{8}$ in 18 months?

$$5 : 100 :: 16.875 : 1.5 \times x = \$225. \quad Ans.$$

$$16.875 \times 12 \div 18 \times .05 = \$225. \quad Ans.$$

When partial payments have been made.

RULE. — Find the amount (sum of the principal and interest) up to the time of the first payment, and deduct the payment therefrom; then find the interest on the remainder up to the next payment, add it to the remainder, or new principal, and from the sum subtract the next payment; and so on for all the payments; then find the amount up to the time of final payment for the final amount.

COMPOUND INTEREST.

If we calculate the interest on a debt for one year, and then on the same debt for another year, and again on the same debt for still another year, the sum will be the *simple* interest on the debt for three years. But, on the contrary, if we calculate the interest on the debt for one year, and then on the *amount* (sum of the principal and interest) for the next year, and then on the second amount for the third year, the sum of the interest so calculated will be the *compound* interest, or yearly compound interest, on the debt for three years; equal to the simple interest on the debt for three years, plus the yearly compound interest on the first year's interest for two years, plus the simple interest on the second year's interest for one year. So, if we divide the time into shorter periods than a year, and proceed for the interest as last suggested, the interest will be compound. Thus we have half-yearly compound interest, or compound interest semi-annually, quarter-yearly compound interest, or compound interest quarterly, &c.

This method of computing interest is predicated upon the natural idea, that interest, when it becomes due by stipulation and is withheld, commences to draw interest, and continues at use to the holder, at the same rate as the principal, until it is paid, like other over-due demands; and that the interest so made matures and becomes due as often, and at the same periods, as that on the principal.

It will be perceived by the foregoing that the *working-time* in compound interest is the interval between the stipulated payments of the interest, or between one stipulated payment of the interest and that of another; and that the *working-rate* is pro rata to the rate per annum.

Thus the *amount* of $100 at semi-annual compound interest for 2 years, at 6 per cent. per annum, is

COMPOUND INTEREST.

$100 \times (1.03)^4 = \$112.550881 = \$112.55$, or

```
       100.
        .03
       ----
        3.
       100.
       ----
       103.
        .03
       ----
        3.09
       103.
       ------
       106.09
          .03
       ------
        3.1827
       106.09
       --------
       109.2727
           .03
       --------
        3.278181
       109.2727
       ----------
```
$112.550881, as before.

If we let P = principal or debt at interest,
r = working-rate of interest,
n = number of intervals into which the whole time is divided for the payment of interest, or number of consecutive intervals for the payment of interest that have transpired without a payment having been made,
i = compound interest,
$A = P + i$ or amount, then

$$A = P(1+r)^n;\ P = \frac{A}{(1+r)^n};\ r = \sqrt[n]{\frac{A}{P}} - 1;$$

$$\frac{A}{P} = (1+r)^n;\ i = A - P.$$

EXAMPLE. — What is the compound interest, or yearly compound interest, on $100 for $1\frac{1}{2}$ years, at 6 per cent. a year?

$100 \times 1.06 \times 1.03 = 109.18 - 100 = \$9.18$. *Ans.*

EXAMPLE. — What is the amount of $560.46, at 7 per cent. compound interest per year, for 6 years and 57 days?

$$560.46 \times (1.07)^6 \times \left(1 + \frac{.07 \times 57}{365}\right) = \$850.29.\ \textit{Ans.}$$

COMPOUND INTEREST.

EXAMPLE. — The *principal* is $250, the *rate* 8 per cent. a year, the *time* 2 years, and the *interest* compound per quarter year: required the *amount*.

$$250 \times \left(1.\frac{.08}{4}\right)^8 = \$292.91. \ Ans.$$

When Partial Payments have been made.

RULE. — Find the amount up to the first payment, and deduct the payment therefrom; then find the amount up to the next payment, and therefrom deduct that payment; and so on for all the payments; then find the amount up to the time of final payment, for the final amount.

EXAMPLE. — A note of hand for $500 and interest from date, at 6 per cent. a year, has been paid in part as follows; viz., two years and four months from the date of the note, by an indorsement of $50; and three years from that indorsement, by an indorsement of $150. It is now eight months since the last payment was made, and the demand is to be settled in full: required the amount at the present time, interest being compound per year.

$$500 \times (1.06)^2 \times 1.02 - 50 = 523.036$$
$$\underline{(1.06)^3}$$
$$622.944$$
$$\underline{150}$$
$$472.944$$
$$\underline{1.04}$$
$$\$491.86. \ Ans.$$

The following table shows $(1 + r)$ raised to all the integer powers from 1 to 12 inclusive; r being taken at 4, 5, 6, 7, 8, and 10 per cent. If the numbers in the column headed years are taken to represent years, then 4 per cent., 5 per cent., &c., at the head of the columns of powers, will stand for per cent. per annum: if they are taken to represent half-years, then 4 per cent., 5 per cent., &c., will stand for per cent. per half-year, &c. The quantities in the columns are powers of $(1 + r)$, of which the numbers referred to and standing opposite, respectively, are the exponents. Thus, 1.26248, in the 6 per cent. column, and against 4 in the column marked years, $= (1.06)^4$; and so with the others. The powers or quantities in the columns are co-efficients in the calculations.

COMPOUND INTEREST.

Years.	4 per cent.	5 per cent.	6 per cent.	7 per cent.	8 per cent.	10 per cent.
1	1.04	1.05	1.06	1.07	1.08	1.10
2	1.0816	1.1025	1.1236	1.1449	1.1664	1.21
3	1.12486	1.15762	1.19102	1.22504	1.25971	1.331
4	1.16986	1.21551	1.26248	1.3108	1.36049	1.4641
5	1.21665	1.27628	1.33823	1.40255	1.46933	1.61051
6	1.26532	1.3401	1.41852	1.50073	1.58687	1.77156
7	1.31593	1.4071	1.50363	1.60578	1.71382	1.94872
8	1.36857	1.47746	1.59385	1.71819	1.85093	2.14359
9	1.42331	1.55133	1.68948	1.83846	1.999	2.35795
10	1.48024	1.62889	1.79085	1.96715	2.15892	2.59374
11	1.53945	1.71034	1.8983	2.10485	2.33164	2.85312
12	1.60103	1.79586	2.0122	2.25219	2.51817	3.13843

NOTE.—If a co-efficient is wanted for a greater number of years or intervals of time than is given in the table, square the tabular co-efficient opposite half that number of intervals, or cube the tabular co-efficient opposite one-third that number of intervals, &c., for the co-efficient required. Thus,

$$1.999^2 = 1.58687^3 = 1.08^{12} \times 1.08^6 = 1.08^{18} = 3.996,$$

the co-efficient for 18 years or intervals at 8 per cent. per interval, &c.

If the compound interest alone is sought on a given principal, subtract 1 from the tabular power corresponding to the time and rate, and multiply the remainder by the given principal; the product will be the compound interest. Thus $(1.26532 - 1) \times 100 = \$26.532$, the yearly compound interest, at 4 per cent. per annum, on $100 for 6 years, or the half-yearly compound interest, at 8 per cent. per annum, on $100 for 3 years, or the half-yearly compound interest, at 4 per cent. per half year, on $100 for 6 half-years.

EXAMPLE.—What is the amount of $125.54, at 5 per cent. compound interest, for 7 years, 21 days?

$1 + \dfrac{21 \times .05}{365} = 1.00288$, the co-efficient for the odd days; and, turning to the 5 per cent. column in the table, we find against 7, in the column of years, 1.4071, the co-efficient for 7 years: then

$$125.54 \times 1.4071 \times 1.00288 = \$178.20. \quad Ans.$$

EXAMPLE.—In what time, at 7 per cent. compound interest per annum, will $1000 gain $462? $A \div P = (1+r)^n$: then $1462 \div 1000 = 1.462$, the co-efficient demanded. Turning now to the 7 per cent. column in the table, we find the nearest less co-efficient there (there being none that exactly corresponds) to be that for 5 years; viz., 1.40255. And $\left(\dfrac{1.462}{1.40255} - 1\right) \div .07 = .60553$, the fraction of a year over 5 years to the answer.

$.60553 \times 365 = 221$ days: 5 years, 221 days. *Ans.*

11*

COMPOUND INTEREST.

The following TABLE is of the same nature as the preceding, and is applicable when the interest becomes due at regular intervals short of a year, or when the working-rate in compound interest is less than 4 per cent.

The quantities in the $1\frac{3}{4}$ per cent. column apply to quarter-yearly compound interest when the rate is 7 per cent. a year; and those in the $1\frac{1}{4}$ per cent. column, to quarterly compound interest when the rate is 5 per cent. a year; also the former are applicable to monthly compound interest at 21 per cent. per annum, and the latter to monthly compound interest at 15 per cent. per annum; and so relatively, throughout the table.

Times.	$3\frac{1}{2}$ per cent.	3 per cent.	$2\frac{1}{2}$ per cent.	2 per cent.	$1\frac{3}{4}$ per cent.	$1\frac{1}{2}$ per cent.	$1\frac{1}{4}$ per cent.	1 per cent.	$\frac{1}{2}$ per cent.
1	1.035	1.03	1.025	1.02	1.0175	1.015	1.0125	1.01	1.005
2	1.07123	1.0609	1.05063	1.0404	1.03531	1.03023	1.02516	1.0201	1.01003
3	1.10872	1.09273	1.07689	1.06121	1.05342	1.04568	1.03797	1.0303	1.01508
4	1.14752	1.12551	1.10381	1.08243	1.07186	1.06136	1.05095	1.0406	1.02015
5	1.18769	1.15927	1.13141	1.10408	1.09062	1.07728	1.06408	1.05101	1.02525
6	1.22925	1.19405	1.15969	1.12616	1.1077	1.09344	1.0774	1.06152	1.03038
7	1.27228	1.22987	1.18869	1.14869	1.12709	1.10984	1.09087	1.07214	1.03553
8	1.31681	1.26677	1.2184	1.17166	1.14681	1.12649	1.10451	1.08286	1.04071
9	1.3629	1.30477	1.24886	1.19509	1.16688	1.14339	1.11831	1.09369	1.04591
10	1.4106	1.34392	1.28008	1.21899	1.1873	1.16054	1.13229	1.10462	1.05114
11	1.45997	1.38423	1.31209	1.24337	1.20808	1.17795	1.14645	1.11567	1.05604
12	1.51107	1.42576	1.34489	1.26824	1.22922	1.19562	1.16078	1.12683	1.06168

EXAMPLE. — What is the amount of $750 for 4 years and 40 days, allowing half-yearly compound interest, at 7 per cent. a year?

In this case, the working-rate for the full periods of time is $3\frac{1}{2}$ per cent., and there are 8 such full periods; then, seeking the co-efficient in the $3\frac{1}{2}$ per cent. column, we find against 8, in the column of times, the quantity or co-efficient 1.31681; and $1 + \dfrac{40 \times .07}{365} =$ 1.00767: therefore

$$750 \times 1.31681 \times 1.00767 = \$995.18. \quad Ans.$$

EXAMPLE. — What is the amount of $1000 at compound interest per quarter-year, at $1\frac{1}{2}$ per cent. per quarter-year, for $4\frac{1}{4}$ years?

$$1000 \times 1.12649^2 \times 1.015 = \$1288.01. \quad Ans.$$

BANK INTEREST OR BANK DISCOUNT.

A bank loans money on a promissory note made payable without interest at a future period. The operation is called *discounting* the note at bank, and is as follows: The bank takes the note, finds the interest on it for three days more time than by its own tenor it has to run, subtracts it from the principal, and hands the balance, called the *avails* of the note, in its own bills, to the party soliciting the loan, or offering the note for discount, as it is called; whereby the note becomes the property of the bank, and the maker and indorsers are held for its payment when it matures.

The three days mentioned are called *days of grace*, and the note does not become due to the bank until three days after it becomes due by its own tenor. These proceedings are sanctioned by usage, and protected by law.

Bank interest, then, is bank discount, and bank discount is bank interest. But bank discount is not *discount*, nor is it what is called *legal* interest on the money loaned. It is the interest on the money loaned, plus the interest on the interest of the loan, plus the interest on the difference of the sum taken and the interest on the loan for the time of the loan! A kind of interest more onerous, if any description of interest be onerous, than compound interest, rate for rate and time for time, as may be readily perceived.

Let $P =$ principal or face of the note.
  $r =$ working-rate of the interest for the time of the loan.
  $a =$ avails of the note or sum borrowed.
  $i =$ bank interest.
  $t =$ time of the loan.
$R : r :: T : t$. R being the rate per cent. per annum, and T one year.
$P = a \div (1 - r)$. $a = P - Pr$. $i = Pr$. $r = (P - a) \div P$.
If we let n represent the time of the note in months,
$r = \dfrac{Rn}{12} + \dfrac{3R}{365}$. But it is the practice with many banks to count the days of grace as so many 360ths of a year.

Putting d to represent the time of the note in days,
$$r = \frac{Rd + 3R}{365}, \text{ true time and rate.}$$

With some banks, it is the practice, in calculating interest, to take the time, when it does not exceed 93 days, as so many 360ths of a year.

A note having 3 months to run from Aug. 10, for instance, will fall due Nov. 10–13; but one having 90 days to run from Aug. 10 will fall Nov. 8–11. The time including grace of the former is 3 mo. 3 ds., and that of the latter 3 mo. 2 ds., mean time. Nevertheless, the former embraces 95 days, or one day more than mean time, and the latter but 93 days.

The following table shows $1-r$, mean time, for the intervals of time set down in the left-hand column; R being taken at 4, 5, 6, 7, and 8 per cent. per annum, as set down at the top of the columns.

Time.		4	5	6	7	8
mo.	ds.	per cent.	per cent.	per cent.	per cent.	per cent.
1	3	.996333	.995417	.9945	.993583	.992667
2	3	.993	.99125	.9895	.98775	.986
3	3	.989667	.987083	.9845	.981917	.979333
4	3	.986333	.982917	.9795	.976083	.972667
5	3	.983	.97875	.9745	.97025	.966
6	3	.979667	.974583	.9695	.964417	.959333
7	3	.976333	.970417	.9645	.958583	.952667
8	3	.973	.96625	.9595	.95275	.946
9	3	.969667	.962083	.9545	.946917	.939333
10	3	.966333	.957917	.9495	.941083	.932667
11	3	.963	.95375	.9445	.93525	.926
12	3	.959667	.949583	.9395	.929417	.919333

Putting k to represent the tabular quantity $1-r$,
$$a = Pk,\ P = a \div k,\ i = P - a = P - Pk.$$

EXAMPLE. — What will be the avails of a note for $1,250 payable in 4 months if discounted at a bank, interest being 7 per cent. a year?

The tabular constant $1-r$, in the 7 per cent. column, against 4 months and 3 days in the time column, is .976083, and
$$\$1,250 \times .976083 = \$1,220.10.\quad Ans.$$

EXAMPLE. — For what sum must I make a note having 6 months to run, in order that the avails at bank, if discounted on the day of the date of the note, may amount to $956.38, interest being 6 per cent. per annum?

By the table, $956.38 \div .9695 = \$986.47.\quad Ans.$

EXAMPLE. — What is the rate of bank interest when the nominal or legal rate is 7 per cent.?

$.07 \div (1 - .07) = .07527 = 7\frac{1}{2} + \frac{27}{1000}$ per cent. $Ans.$

NOTE. — A note having 5 months to run from Feb. 1 will fall due July 1–4; and the time, including grace, is 5 mo. 3 ds. = 155 days, mean time. But the time in days from Feb. 1 to July 4, when February has but 28 days, is 153 days only, or 2 days short of mean time.

DISCOUNT.

Discount is a deduction of the interest on the present worth or availability of a debt not yet due, in consideration of its present payment. The *principal* is the present nominal value of the debt, interest included, if any interest has accrued. The *time* is the interval from the present to the date at which the debt will become due. The *rate* is the legal rate of interest, if no other rate is specified; and the *present worth* is that sum of money, which, if put at interest at the same rate and for the same time as the discount, will amount to the principal.

Let a represent the principal, d the discount, w the present worth, and i the interest on one dollar for the time and at the rate of the discount.

$$w = a \div (1+i) = a - d. \quad d = ai \div (1+i) = a - w.$$
$$a = d(1+i) \div i = d + w.$$

Example. — Required the discount on $250 for 8 months at 6 per cent.

The interest on $1 for 8 months at 6 per cent. is .04 of a dollar, or 4 cts.; and

$$250 \times .04 \div (1 + .04) = \$9.6154. \quad Ans.$$

Example. — Required the present worth of $1272.62 due 247 days hence, discount 7 per cent.

The interest on $1 for 247 days at 7 per cent. $= 247 \times .07 \div 365 = 0.04737$, and

$$1272.62 \div 1.04737 = \$1215.06. \quad Ans.$$

Note.—"*Taking off*, in common parlance, a certain per centum from the face of a demand, is equal to deducting the interest, at that rate per centtum, on the present worth for 1 year, plus the interest on the interest of the present worth, at the same rate per centum for 1 year.

COMPOUND DISCOUNT.

Compound Discount is to compound interest what simple discount is to simple interest. In both cases of discount, the difference between the principal and the discount is that sum of money, which, if put at interest for the same length of time, at the same rate, and in the same general manner as the discount, will amount to the principal.

Rule. — Add 1 to the rate per cent. of the discount for the

COMPOUND DISCOUNT.

working-time, and raise the sum to a power corresponding with the number of working-times; divide the principal by the power, and the quotient will be the present worth; subtract the present worth from the principal, and the remainder will be the compound discount.

NOTE.—The TABLES of the powers of $1+r$, applicable to compound interest, are equally applicable to compound discount.

EXAMPLE.— Required the present worth of a debt of $250, allowing yearly compound discount, at 7 per cent. a year, for 3 years 84 days.

$$1 + \frac{.07 \times 84}{365} = 1.01611,$$ the working-rate for the 84 days, and

$$250 \div (1.07^3 \times 1.01611) = \$200.84. \quad Ans.$$

EXAMPLE. — What is the present worth of a debt of $150.25, due 3 years, 3 months, and 10 days hence, without interest, allowing compound discount per quarter-year, at $1\frac{1}{2}$ per cent. per quarter-year?

$$150.25 \div \left(1.015^{13} \times 1. \frac{.06 \times 10}{365}\right) = Ans.$$

By table, $150.25 \div (1.19562 \times 1.015 \times 1.00164) =$
$\$123.61. \quad Ans.$

NOTE.—What is here denominated the debt, or principal, represents the debt at the close of the time of the discount; that is, if the debt be on interest, the interest must be included in what is here called the debt, or principal.

PROFIT AND LOSS.

The term "PROFIT AND LOSS," as intimated in treating of PERCENTAGE, relates to the positive and negative interests in business, and embraces the idea of both.

Both profit and loss are absolute quantities, and are expressed by the difference of the cost price and selling price that limit them. They are usually, however, estimated by percentage, predicated upon the first-mentioned price or prime cost.

When the selling price is greater than the cost price, or when the money obtained by the disposal of property exceeds what the property cost, the difference is positive, and denotes increase, profit, or gain. Conversely, when the cost price is greater than the selling price, or when property is disposed of for less money than it cost, the difference is negative, and denotes decrease, loss, or

PROFIT AND LOSS.

waste. So, the difference of the two prices, divided by the cost price, expresses the rate of gain on the cost when the selling price is the greater,—expresses the rate of loss on the cost when the cost price is the greater.

Let c represent the cost price, purchase price, par value, or sum of money paid for the property; s, the selling price, trade price, premium price, or sum of money received in exchange for the property; r, the rate of the profit or loss; p, the rate per cent. of the profit or loss.

To find the rate or rate per cent. of the profit or loss.

$r = \dfrac{s \sim c}{c}$. $p = \dfrac{(s \sim c)\,100}{c}$. Moreover, when the difference is positive, $r = \dfrac{s}{c} - 1$; and, when it is negative, $r = 1 - \dfrac{s}{c}$.

EXAMPLE.— Paid $4 for an article, and sold it for $5. What per cent. was gained? 5 is more than 4 by what per cent. of 4? The difference of 5 and 4 is what per cent. of 4? $5 - 4 = \$1$, gained; and $\dfrac{5 \sim 4}{4} = .25 = \frac{1}{4} - 1$. 25 per cent. *Ans.*

EXAMPLE.— Paid $5 for an article, and sold it for $4. What per cent. was lost? 4 is less than 5 by what per cent. of 5? The difference of 4 and 5 is what per cent. of 5? $4 - 5 = -1 = \$1$, lost; and $\dfrac{5 \sim 4}{5} = .20 = 1 - \frac{4}{5}$. 20 per cent. *Ans.*

EXAMPLE.— A whistle that cost 3 cents was sold for 20 cents! The profit was how much per cent? $(20 \sim 3) \div 3 = 5\frac{2}{3}$ or $566\frac{2}{3}$ per cent. *Ans.*

EXAMPLE.— A fop paid $10 for a well-made and well-fitting pair of boots for his own wear, that were worth what they cost him; but, being told that they were unfashionably large, sold them for $4. His vanity cost him what per cent. of the purchase price? $1 - \frac{4}{10} = .6$ or 60 per cent. *Ans.*

To find a price long a given per cent. of the cost, or to find a selling price that shall be the sum of the cost price and a given per cent. of it.

$s = c + cr = c\,(1 + r) = c\,(100 + p) \div 100.$

EXAMPLE.— At what price must I sell an article that cost $2.35 to gain 25 per cent.? 2.35, more 25 per cent. of it, is how much? The sum of $2.35 and 25 per cent. of it is how much? $2.35 + 2.35 \times .25 = 2.35 \times 1.25 = \$2.93\frac{3}{4}$. *Ans.*

To find a price short a given per cent. of the cost, or to find a selling price that shall be the difference of the cost price and a given per cent. of it.

$$s = c - cr = c(1-r) = c(100-p) \div 100.$$

EXAMPLE. — I have a damaged article of merchandise that cost $2.75, and I wish to mark it for sale at 30 per cent. below cost. At what price shall I mark it? 2.75 less 30 per cent. of it is how much? The difference of $2.75 and 30 per cent. of it is how much? 2.75 (1 — .30) = 2.75 × .7 = $1.925. *Ans.*

To find the cost price when the selling price and profit per cent. are given.

$$s = c + cr = c(1+r) \therefore c = s \div (1+r) = 100s \div (100+p).$$

EXAMPLE. — What cost that article whose selling price, $4, is long 25 per cent. of the cost? What price, more 25 per cent. of it, is equal to $4? $4 is the sum of what price and 25 per cent. of it? 400 ÷ 125 = $3.20. *Ans.*

To find the cost price when the selling price and loss per cent. are given.

$$s = c - cr = c(1-r) \therefore c = s \div (1-r) = 100s \div (100-p)$$

EXAMPLE. — What cost that article whose selling price, $375, is short 7 per cent. of the cost? What price less 7 per cent. of it is equal to $375? $375 is the difference of what price and 7 per cent. of it?

375 ÷ (1 — .07) = 375 ÷ .93 = 375 × 100 ÷ (100 — 7) = $403.226. *Ans.*

EQUATION OF PAYMENTS, OR AVERAGE.

AVERAGE consists in finding the time at which several sums, falling due at different dates, become due if taken collectively.

RULE. — Multiply each sum respectively by the number of days it falls due later than that falling due at the earliest date, and divide the sum of the products by the sum of the several sums. The quotient will be the number of days subsequent to the earliest date at which the whole will mature, or averages due.

NOTE. — AVERAGE gives no "*interest on interest*" to the creditor. It does not give him his just due. It estimates by way of the *interest* on both sides, on the sums falling due prior to the average date, and on those falling due subsequently, and not by the interest on those falling due prior, and by the *discount* on those falling due subsequent, as would be strictly correct. The practice is against the creditor or holder of the demands, in like manner and relative extent, as shown in note under DISCOUNT.

EQUATION OF PAYMENTS.

The following exhibits the face of an account in the ledger, and the time (date) at which it averages due is required.

1860, April 10 —— $250.26 — 6 mo. Due Oct. 10.
" June 25 —— 320.56 — 6 "   " Dec. 25.
" July 10 —— 50.62 — 3 "   " Oct. 10.
" Aug. 1 —— 210.84 — 4 "   " Dec. 1.
"   " 18 —— 73.40 — 5 "   " Jan. 18.
" Oct. 15 —— 100. — cash   " Oct. 15.

EXAMPLE. — Practical method of stating and working.
1860. Due Oct. 10, $301
"   " Dec. 25,   321 × 76 = 24396.
" · "   "   1,   211 × 52 = 10972.
"   " Jan. 18,   73 × 100 = 7300.
"   " Oct. 15,   100 × 5 =   500.
                 ────   ─────
                 1006   ) 43168 (43 days, = Nov. 22, 1860.
                                              Ans.

COMPOUND AVERAGE.

COMPOUND AVERAGE consists in finding the time at which the *balance* of an account or demand averages due, whose sides — the debit and the credit — average due at different dates.

RULE. — Multiply the less sum or side by the difference in days between the two dates — that at which the debit side averages due and that at which the credit side averages due — and divide the product by the difference of the sums or sides; the quotient will be the number of days that one of the dates must be set back, or the other forward, to mark the time sought; for which last,

SPECIAL RULE.

Earlier date with larger sum, set back from earlier.
Later date with larger sum, set forward from later.

EXAMPLE. — The debit side of an account in the ledger foots up $400, and averages due Oct. 12, 1860; the credit side of the same account foots $300, and averages due Nov. 16, 1860. At what date does the balance or difference between the two sides average due?

400   300
300   35
───   ───
100) 10500 (105 days earlier than Oct. 12, = June 29, 1860. *Ans.*

EXAMPLE. — The debit side of an obligation foots $250, and averages due May 17, 1860; the credit side of the same obligation foots $175, and averages due May 1, 1860. At what date does the difference of the sides average due?

250   175
175   16
───   ───
75) 2800 (37½ days later than May 17, = June 23, 1860. *Ans.*

GENERAL AVERAGE.

It is the established usage that whatever of either of the three commercial interests — the ship, the cargo, or the freight — is voluntarily sacrificed or destroyed for the general good, or with the view of saving the most that may be saved when all is in imminent danger of being lost, is matter of general loss to the respective interests, and not more especially to the interest voluntarily abandoned than to the others. So, too, the losses and damages incident to the voluntary sacrifice, and collateral therewith, together with the expenditures which the master has been compelled to make for the general good, in consequence of disaster, are matters of general average, or are to be contributed for, *pro rata*, by the several interests.

The contributory interests are the ship, the cargo, and the freight, at their net values, independent of charges, premiums paid for insurance, &c.

The contributory value of the ship, generally, is her value at the port of departure at the time of leaving, less the premium paid for her insurance.

The contributory value of the cargo is its net value, in a sound state, at the port of destination, if the voyage be completed; or its invoice value if the voyage be broken up and the cargo returned to the port whence it was shipped; or its market-value at any intermediate port, where of necessity it is discharged and disposed of. The value of the goods jettisoned, and to be contributed for, is their value after the same manner; and that value is a part of the contributory value of the cargo, as well as a matter of general average.

The contributory value of the freight, generally, is the gross amount or amount per freight-list, less one-third part thereof, in most of the States; but, in the State of New York, one-half thereof, for seamen's wages and other expenses. The loss of freight by jettison, when any freight is earned, is matter of general average. If the cargo is transshipped on board another vessel, and in that way sent to the port of destination, the contributory value of the freight is the gross amount, less the sum paid the other vessel.

The voluntary damage to the ship, with a view to the general good, — such as throwing over her furniture, destroying her equipments, cutting away her masts, breaking up her decks to get at the cargo for the purpose of throwing it over, &c., — is contributed for at two-thirds the cost of repairing and restoring; the new articles being supposed one-half better, or worth one-half more, than the old.

GENERAL AVERAGE. 135

If we let $V =$ contributory value of the vessel,
$C =$ contributory value of the cargo,
$F =$ contributory value of the freight,
$d =$ aggregate amount of losses to be averaged, then $d \div (V + C + F) = r$, the per cent. of each interest that each must contribute, and

$V \times r =$ Vessel's share of the loss,
$C \times r =$ Cargo's share of the loss,
$F \times r =$ Freight's share of the loss.

When a contributory interest's share of the loss is to be distributed among the several owners of that interest, the same *pro rata* method is to be observed: thus

$A \times r =$ sum A must contribute,
$B \times r =$ sum B must contribute,
$D \times r =$ sum D must contribute;

A, B, and D being A's, B's, and D's respective shares in that interest.

ASSESSMENT OF TAXES.

G = amount of taxable property, real and personal, as per grand list.
A = amount of money to be raised, including the whole poll-tax.
T = amount of money to be raised on property alone.
n = number of ratable polls.
h = poll-tax per head.
r = rate per cent. to be raised on taxable property.
P = an individual's taxable property, as per grand list.
b = P's poll-tax.

$T = A - hn.$ $r = T \div G.$ $Pr + b =$ P's tax, including poll.

INSURANCE.

INSURANCE is a written contract of indemnity, called the *policy*, by which one party (the *insurer* or *underwriter*) engages, for a stipulated sum, called the *premium* (usually a per cent. on the value of the property insured), to insure another against a risk or loss to which he is exposed.
Let P = Principal, or amount insured on,
 r = rate per cent. of insurance,
 a = premium for insurance.

$$a = Pr. \quad r = a \div P. \quad P = a \div r.$$

EXAMPLE. — What is the premium for insuring on $4500 at 1½ per cent.?

$4500 \times .015 = \$67.50.$ *Ans.*

LIFE-INSURANCE.

Life-insurance is predicated upon the even chance in years, called the *expectation of life*, that an individual in general health at any given age appears by the rates of mortality to have of living beyond that age.

The Carlisle Tables of Expectation, column C in the following tables, are used almost or quite exclusively in England, and by some insurance-companies in the United States; while those by Dr. Wigglesworth, column W, computed with special reference to the rates of mortality in this country, are used by others.

The Supreme Court of Massachusetts has adopted the Wigglesworth rates of expectation in estimating the value of life-annuities and life-estates.

TABLE

Of Ages and Expectations from Birth to 103 Years.

Age	C.	W.	Age	C.	W.	Age	C.	W.	Age	C.	W.
0	38.72	28.15	26	37.14	31.93	52	19.68	20.05	78	6.12	6.59
1	44.68	36.78	27	36.41	31.50	53	18.97	19.46	79	5.80	6.21
2	47.55	38.74	28	35.69	31.08	54	18.28	18.92	80	5.51	5.85
3	49.82	40.01	29	35.00	30.66	55	17.58	18.35	81	5.21	5.50
4	50.76	40.73	30	34.34	30.25	56	16.89	17.78	82	4.93	5.16
5	51.25	40.88	31	33.68	29.83	57	16.21	17.20	83	4.65	4.87
6	51.17	40.69	32	33.03	29.43	58	15.55	16.63	84	4.39	4.66
7	50.80	40.47	33	32.36	29.02	59	14.92	16.04	85	4.12	4.57
8	50.24	40.14	34	31.68	28.62	60	14.34	15.45	86	3.90	4.21
9	49.57	39.72	35	31.00	28.22	61	13.82	14.86	87	3.71	3.90
10	48.82	39.23	36	30.32	27.78	62	13.31	14.26	88	3.59	3.67
11	48.04	38.64	37	29.64	27.34	63	12.81	13.66	89	3.47	3.56
12	47.27	38.02	38	28.96	26.91	64	12.30	13.05	90	3.28	3.43
13	46.51	37.41	39	28.28	26.47	65	11.79	12.43	91	3.26	3.32
14	45.75	36.79	40	27.61	26.04	66	11.27	11.96	92	3.37	3.12
15	45.00	36.17	41	26.97	25.61	67	10.75	11.48	93	3.48	2.40
16	44.27	35.76	42	26.34	25.19	68	10.23	11.01	94	3.53	1.98
17	43.57	35.37	43	25.71	24.77	69	9.70	10.50	95	3.53	1.62
18	42.87	34.98	44	25.09	24.35	70	9.18	10.06	96	3.46	
19	42.17	34.59	45	24.46	23.92	71	8.65	9.60	97	3.28	
20	41.46	34.22	46	23.82	23.37	72	8.16	9.14	98	3.07	
21	40.75	33.84	47	23.17	22.83	73	7.72	8.69	99	2.77	
22	40.04	33.46	48	22.50	22.27	74	7.33	8.25	100	2.28	
23	39.31	33.08	49	21.81	21.72	75	7.01	7.83	101	1.79	
24	38.59	32.70	50	21.11	21.17	76	6.69	7.40	102	1.30	
25	37.86	32.33	51	20.39	20.61	77	6.40	6.99	103	0.83	

Thus, by the tables, a man in general good health at 21 years of age has an even chance, by the Carlisle rate of mortality, of living 40¾ years longer; by the Wigglesworth rate, of living 33$\frac{84}{100}$ years longer. So a man in general good health, at 60 years of age, has, by the Carlisle rate, an even chance of living 14.34 years longer; by the Wigglesworth rate, an even chance of living 15.45 years longer, etc.

FELLOWSHIP.

FELLOWSHIP calls for the distribution of a given effect to each of the several causes associated in its production, proportional to their respective magnitudes one with another.

It is a rule, therefore, adapted to the use of partners associated in business, in achieving a *pro rata* distribution among themselves as individuals, of the profits or losses pertaining to the company.

RULE. — Multiply each partner's investment or share of the capital stock, by the whole gain or loss, and divide the product by the sum of all the shares, or gross capital.

EXAMPLE. — Three men, A, B, and C, enter into partnership. A invests $500, B $700, and C $300. They trade and gain $400. What is each partner's share of the profits?

A, $500	$500 \times 400 \div 1500 =$	$133.33\frac{1}{3} =$ A's share.
B, 700	$700 \times 400 \div 1500 =$	$186.66\frac{2}{3} =$ B's "
C, 300	$300 \times 400 \div 1500 =$	$80.00 =$ C's "

$1500 =$ gross capital. $400.00 Proof.

EXAMPLE. — D's investment of $600 has been employed eight months; E's, of $500, five months; and F's, of $300, five months; the profits of the company are $500, and are to be divided *pro rata* among the partners. What is each partner's share?

D, $600 \times 8 = 4800 \times 500 \div 8800 = $272.73,$ D's share.
E, $500 \times 5 = 2500 \times 500 \div 8800 = 142.05,$ E's "
F, $300 \times 5 = 1500 \times 500 \div 8800 = 85.22,$ F's "
                         $\overline{8800}$ $\overline{\$500.}$ Proof.

EXAMPLE. — Of $120 distributed, there were given to A, $\frac{1}{3}$; to B, $\frac{1}{4}$; to C, $\frac{1}{5}$; and to D, $\frac{1}{6}$, and there was nothing remaining. What sum did each receive?

$\frac{1}{3}$ of $120 = 40 \times 120 \div 114 = \$42\frac{2}{19} =$ A's share.
$\frac{1}{4}$ of $120 = 30 \times 120 \div 114 = 31\frac{11}{19} =$ B's "
$\frac{1}{5}$ of $120 = 24 \times 120 \div 114 = 25\frac{5}{19} =$ C's "
$\frac{1}{6}$ of $120 = 20 \times 120 \div 114 = 21\frac{1}{19} =$ D's "
                        $\overline{114}$ $\overline{\$120.}$ Proof.

EXAMPLE. — Divide the number 180 into 3 parts, which shall be to each other as 2, 3, 4.

$\frac{2}{9}$ of $180 = 90 \times 180 \div 195 = 83.08$
$\frac{3}{9}$ of $180 = 60 \times 180 \div 195 = 55.38$
$\frac{4}{9}$ of $180 = 45 \times 180 \div 195 = 41.54$
                $\overline{195}$ $\overline{180.00}$ Proof.

Example. — $400 are to be divided between A, B, and C, in the ratio of $\frac{1}{2}$ to A, $\frac{1}{2}$ to B, and $\frac{1}{4}$ to C; how much will each receive?

$\frac{1}{2}$ of 400 = 200, and 200 × 400 ÷ 500 = $160 = A's share.
$\frac{1}{2}$ of 400 = 200, and 200 × 400 ÷ 500 = 160 = B's share.
$\frac{1}{4}$ of 400 = 100, and 100 × 400 ÷ 500 =   80 = C's share.
500 $400. Proof.

ALLIGATION.

Alligation *Medial* is a method by which to find the mean price of a mixture or compound, consisting of two or more articles or ingredients, the quantity and price of each being given.

Rule. — Multiply each quantity by its price, and divide the sum of the products by the sum of the quantities; the quotient will be the price per unity of measure of the mixture; and, having found the price of the given quantities as mixed, any quantities of the same materials, taken in like proportions, will be at the same price.

Example. — If 20 lbs. of sugar at 8 cents, 40 lbs. at 7 cents, and 80 lbs. at 5 cents per pound, be mixed together, what will be the mean price, or price per pound, of the mixture?

20 × 8 = 160
40 × 7 = 280
80 × 5 = 400
140   ) 840 (6 cents. *Ans.*

The several kinds, then, at their respective prices, taken in the proportion of 1 at 8, 2 at 7, and 4 at 5 cts., will form a mixture worth 6 cts. a pound.

Example. — If 10 lbs. of nickel are worth $2, and 24 lbs. of copper are worth 4\frac{1}{2}$, and 8 lbs. of zinc are worth 40 cts., and 1 lb. of lead is worth 5 cts., what are 5 lbs. of *pretty good* German silver worth?

$\frac{(200+450+40+5) \times 5}{43}$ = 81 cents. *Ans.*

Alligation *Alternate* is a method by which to find what quantity of each of two or more articles or ingredients, whose prices or qualities are given, must be taken to form a mixture or compound that shall be at a given price or of a given quality between the two extremes. It also applies to the finding of relative quantities when the quantity of one or more of the articles is limited.

Rule. — Connect the given prices or qualities — a less than the given mean with that one or either one that is greater — and to the extent that all be thus connected; then place the difference between each given and the given mean opposite, not the given, or the given mean, but the given with which it is alligated; the number standing opposite each price or quality will be the quantity that must be taken at that price, or of that quality, to form a mixture or compound at the price or of the quality desired. And, being proportions respectively to each other, they may be taken in ratio greater or less, as desired.

EXAMPLE. — In what proportions shall I mix teas at 48 cents a pound and 54 cents a pound, that the mean price may be 50 cents a pound?

In the proportions

$$50 \begin{Bmatrix} 48 \\ 54 \end{Bmatrix} \begin{Bmatrix} 4 \text{ lbs at 48 cts.} \\ 2 \text{ lbs. at 54 cts.} \end{Bmatrix} Ans.$$

Or, as 2 at 48 to 1 at 54.

Proof. $\begin{Bmatrix} \overline{2 \times 48} + \overline{1 \times 54} = 150. \\ 3 \times 50 = 150. \end{Bmatrix}$

EXAMPLE. — In what proportions shall I mix teas at 48, 54, and 72 cents a pound, that the mixture may average 60 cents a pound?

$$60 \begin{Bmatrix} 48 \\ 54 \\ 72 \end{Bmatrix} \begin{matrix} 12, \\ 12, \\ 12 + 6, \end{matrix} \begin{Bmatrix} 12 \text{ at 48} \\ 12 \text{ at 54} \\ 18 \text{ at 72} \end{Bmatrix} = \begin{Bmatrix} 2 \text{ at 48} \\ 2 \text{ at 54} \\ 3 \text{ at 72} \end{Bmatrix} Ans.$$

EXAMPLE. — A wine dealer has received an order for a quantity of wine at 50 cts. a gallon. He has none ready *manufactured* at that price. He has it at 40 cts., at 56 cts., and at 80 cents a gallon, and he has water that cost him nothing. He wishes to fill the order with a mixture composed of the four materials — the water and the three different priced wines. In what proportions must he mix them, that the mean or average price may be 50 cents?

Ans.   *Ans.*

$$50 \begin{bmatrix} 00 \\ 40 \\ 56 \\ 80 \end{bmatrix} \begin{bmatrix} 6 \\ 30 \\ 50 \\ 10 \end{bmatrix} \quad Or, \ 50 \begin{bmatrix} 00 \\ 40 \\ 56 \\ 80 \end{bmatrix} \begin{bmatrix} 30 & = 30 \\ 6 + 30 = 36 \\ 10 & = 10 \\ 50 + 10 = 60 \end{bmatrix}$$

= 96 gals.   = 136 gals.

Ans.   *Ans.*

$$Or, \ 50 \begin{bmatrix} 00 \\ 40 \\ 56 \\ 80 \end{bmatrix} \begin{bmatrix} 6 + 30 \\ 30 \\ 50 \\ 50 + 10 \end{bmatrix} \quad Or, \ 50 \begin{bmatrix} 00 \\ 40 \\ 56 \\ 80 \end{bmatrix} \begin{bmatrix} 6 \\ 6 + 30 \\ 50 + 10 \\ 10 \end{bmatrix}$$

= 176 gals.   = 112 gals.

If, now, having found the proportions desired, it is wished to limit one of the articles in quantity — say the best wine to 8 gallons in the

INVOLUTION — EVOLUTION.   141

mixture — the proportions of the remaining articles thereto are found thus: —
Instance, 1st example, —

$10 : 8 :: 50 = 40$
$10 : 8 :: 30 = 24$ } And the mixture will consist of
$10 : 8 :: 6 = 4\frac{4}{5}$ } $8 + 40 + 24 + 4\frac{4}{5} = 76\frac{4}{5}$ gallons.

If, instead, it is desired to mix a given quantity, say 100 gallons, and proportioned, say as in first example, the quantity to be taken of each is ascertained by the following

RULE. — As the sum of the relative quantities is to the quantity required, so is each relative quantity to the quantity required of it respectively.
The sum of the relative quantities alluded to is $6 + 30 + 50 + 10 = 96$; then,

$96 : 100 :: 6 = 6\frac{1}{4}$
$96 : 100 :: 30 = 31\frac{1}{4}$
$96 : 100 :: 50 = 52\frac{1}{12}$
$96 : 100 :: 10 = 10\frac{5}{12}$

INVOLUTION.

INVOLUTION consists in involving, that is, in multiplying a number one or more times into itself. The number so involved is called the *root*, and the product arising from such involution, its *power*.

The *second power*, or *square*, of the root, is obtained by multiplying the root *once* into itself, as $4 \times 4 = 16$; 4 being the root and 16 its square.

The *third power*, or *cube*, of a number, is obtained by multiplying the number twice into itself, as $4 \times 4 \times 4 = 64$; and so on for any power whatever.

When a number is to be involved into itself, a small figure called the *index* or *exponent* is placed at its right, indicating the number of times it is to be so involved, or the power to which it is to be raised. Thus, $3^1 = 3 \times 3 \times 3 \times 3 = 81$; and $4^3 = 4 \times 4 \times 4 = 64$.

EVOLUTION.

EVOLUTION is the opposite of Involution. It consists in finding a root of a given number, instead of a power of a given root.

When the root of a number is required or indicated, the number is written with the $\sqrt{}$ before it: and the character or denomination of the root, if it be other than the square root, is defined by an index figure placed over the sign. When the square root of a number is required, the sign ($\sqrt{}$) is placed before the number, but the index (2) is usually omitted. Thus, $\sqrt{25}$, shows that the square root of 25 is required, or to be taken; and $\sqrt[3]{25}$ shows that the cube root is required. The operation is usually called extracting the root.

TO EXTRACT THE SQUARE ROOT.

RULE — 1. Separate the given number into periods of two figures each, by placing a point over the *first* figure, *third*, *fifth*, &c., counting from right to left — the root will consist of as many figures as there are periods.

2. Find the greatest square in the left hand period, and place its root in the quotient; subtract the square of the root from the left hand period, and to the remainder bring down the next period for a dividend.

3. Multiply the root so far found — the figure in the quotient — by 2, for a divisor; see how many times the divisor is contained in the dividend, except the right hand figure, and place the result (the number of times it is contained) in the quotient, to the right of the figure already there, and also to the right of the divisor; multiply the divisor, thus increased, by the last figure in the quotient, and subtract the product from the dividend, and to the remainder bring down the next period for a dividend.

4. Multiply the quotient — the root so far found (now consisting of two figures) — by 2, as before, and take the product for a divisor; see how many times the divisor is contained in the dividend, except the right hand figure, and place the result in the quotient, and to the right of the divisor, as before; multiply the divisor, as it now stands, by the figure last placed in the quotient, and subtract the product from the dividend, and to the remainder bring down the next period for a dividend, as before.

5. Multiply the quotient (now consisting of 3 figures) by 2, as before, and take the product for a divisor, and in all respects proceed as when seeking for the last two figures in the quotient. The quotient, when all the periods have been brought down and divided, will be the root sought.

NOTE. — 1. If there is a remainder after finding the integer of a root, annex periods of ciphers thereto, and proceed as when seeking for the integer. The quotient figures will be the decimal portion of the root.
2. If the given number is a decimal, or consists of a whole number and decimal, point off the decimal from left to right, by placing the point over the *second*, *fourth*, *sixth*, &c., figures therein, and fill the last period, if incomplete, by annexing a cipher.
3. If the dividend does not contain the divisor, a cipher must be placed in the quotient, and also at the right of the divisor, and the next period brought down; then the dividend must be divided by the divisor as increased.
4. If the quotient figure, obtained by dividing by the double of the root, is too large, as will sometimes be the case, (see 3d Example) it must be dropped, and a less — one which is the true measure — taken in its stead.

EVOLUTION. 143

EXAMPLE. — Required the square root of 123456.432.

```
    123456.4320 ( 351.3636+.  Ans.
      9
 65 )  334
       325
    70.)  956
          701
       7023 ) 25543
              21069
          70266 ) 447420
                  421596
             702723 ) 2582400
                      2108169
                7027266 ) 47423100
                          42163596
                           5259504
```

EXAMPLE. — Required the square root of 10621. Also, of 28561.

```
     10621 ( 103.05+.  Ans.    |    28561 ( 169.  Ans.
     1                         |    1
203 ) 00621                    |  26 ) 185
      609                      |       156
  20605 ) 120000               |    329 ) 2961
         103025                |          2961
          16975
```

TO EXTRACT THE CUBE ROOT.

RULE — 1. Separate the given number into periods of three figures each, by placing a point over the *first, fourth, seventh,* &c., counting from right to left — the root will consist of as many figures as there are periods.

2. Find the greatest cube in the left hand period, and place its root in the quotient; subtract the cube of the root from the left hand period, and to the remainder bring down the next period for a dividend.

3. Multiply the square of the quotient by 300, for a divisor; see how many times the divisor is contained in the dividend, and place the result (except that the remainder is large, diminished by one or two units) in the quotient.

4. Multiply the divisor by the figure last placed in the quotient, and to the product add the square of the same figure, multiplied by the other figure, or figures, in the quotient, and by 30 ; and add also thereto

144 EVOLUTION.

the cube of the same figure, and take the sum for the subtrahend; subtract the subtrahend from the dividend, and to the remainder bring down the next period for a dividend, with which proceed as with the preceding, so continuing until the whole is completed.

NOTE—1. Decimals must be pointed from left to right, by placing a point over the *third, sixth*, &c., figures in that direction.
2. If the divisor is not contained by the dividend, place a cipher in the quotient, and annex two ciphers to the divisor, and bring down the next period for a dividend, and use the divisor, as thus increased, for finding the next quotient figure.
3. If there is a remainder after finding the integer of the root, annex a period of three ciphers thereto, and proceed for the decimal of the root as if seeking for the integer, annexing a period of three ciphers to each remainder until the decimal is carried to as many places of figures as desired.

EXAMPLE.—Required the cube root of 47421875.6324.

$$47421875.632400\ (\ 361.959+.$$
$$27 \qquad\qquad\qquad Ans.$$

$$3^2 \times 300 = 2700\ )\ 20421$$
$$6$$

$$\begin{array}{r}16200\\6^2 \times 3 \times 30 = 3240\\6^3 = 216 = 19656\end{array}$$

$$36^2 \times 300 = 388800\ )\ 765875$$
$$1$$

$$\begin{array}{r}388800\\1^2 \times 36 \times 30 = 1080\\1^3 = 1 = 389881\end{array}$$

$$361^2 \times 300 = 390963Q0\ )\ 375994632$$
$$9$$

$$\begin{array}{r}351866700\\9^2 \times 361 \times 30 = 877230\\9^3 = 729 = 352744659\end{array}$$

$$3619^2 \times 300 = 3920148300\ )\ 23249973400$$
$$5$$

$$\begin{array}{r}19645741500\\5^2 \times 3619 \times 30 = 2714250\\5^3 = 125 = 19648455875\end{array}$$

$$36195^2 \times 300 = 393023107500\ )\ 3601517525000$$
$$9$$

$$\begin{array}{r}3537210667500\\9^2 \times 36195 \times 30 = 87953850\\9^3 = 729 = 3537298622079\end{array}$$

$$64218902921$$

EVOLUTION.

EXAMPLE. — Required the cube root of 32768. Also, of 8489664.

$$
\begin{array}{r}
32768(32. \\
27 \quad Ans.\\ \hline
\end{array}
$$
$3^2 \times 300 = 2700$) 5768
2
$\overline{5400}$
$2^2 \times 3 \times 30 = 360$
$2^3 = 8 = 5768$

$$
\begin{array}{r}
8489664(204.\\
8 \quad Ans.\\ \hline
\end{array}
$$
$2^2 \times 300 = 120000$) 489664
4
$\overline{480000}$
$4^2 \times 20 \times 30 = 9600$
$4^3 = 64 = 489664$

General Rule for extracting the roots of all powers, or for finding any proposed root of a given number.

1. Point off the given number into periods of as many figures each, counting from right to left, as correspond with the denomination of the root required; that is, if the cube root be required, into periods of *three* figures, if the fourth root, into periods of *four* figures, &c.

2. Find the first figure of the root by inspection or trial, and place it at the right of the number, in the form of a quotient; raise this quotient figure to a power corresponding with the denomination of the root sought, and subtract that power from the left hand period, and to the remainder bring down the first figure of the next period, for a dividend.

3. Raise the root thus far found (the quotient figure) to a power next inferior in denomination to that of the root required, multiply this power by the number or index figure of the root required, and take the product for a divisor; find the number of times the divisor is contained in the dividend, and place the result (except that the remainder is large, diminished by one or two units) in the quotient, for the second figure of the root.

4. Raise the root thus far found (now consisting of two figures) to a power corresponding in denomination with the root required, and subtract that power from the two left hand periods, and to the remainder bring down the first figure of the third period, for a dividend; find a new divisor, as before, and so proceed until the whole root is extracted.

EXAMPLE.— Required the fifth root of 45435424.

$$
\begin{array}{r}
45435424(34. \quad Ans.\\
3^5 = 243. \\ \hline
\end{array}
$$
$3^4 \times 5$) 2113
$34^5 = 45435424$
$\ldots\ldots\ldots$

EXAMPLE. — Required the fifth root of 432040.0554.

$$
\begin{array}{l}
\quad\quad 43204\overset{.}{0}.0354\overset{.}{0}\ (\ 13.4 +.\quad Ans. \\
1^5 = 1 \\
1^4 \times 5\ )\ \overline{33} \\
\quad 13^5 = 371293 \\
13^4 \times 5\ )\ \overline{607470} \\
\quad 13.4^5 = 43204003424 \\
\quad\quad\quad\quad\quad \overline{\cdots\cdots 116}
\end{array}
$$

For instructions touching special cases, see NOTES relative to the extraction of the square root, and to the extraction of the cube root.

The $\sqrt{}$ of the $\sqrt{}$ of any number $=$ $\sqrt[4]{}$ of that number
" $\sqrt{}$ of the $\sqrt[3]{}$ $=$ $\sqrt[6]{}$.
" $\sqrt{}$ of the $\sqrt{}$ of the $\sqrt{}$ $=$ $\sqrt[8]{}$.
" $\sqrt[3]{}$ of the $\sqrt[3]{}$ $=$ $\sqrt[9]{}$.
" $\sqrt{}$ of the $\sqrt[5]{}$ $=$ $\sqrt[10]{}$, &c.

ARITHMETICAL PROGRESSION.

A series of three or more numbers, increasing or decreasing by equal differences, is called an *arithmetical progression*. If the numbers progressively increase, the series is called an *ascending arithmetical progression;* and if they progressively decrease, the series is called a *descending arithmetical progression.*

The numbers forming the series are called the *terms* of the progression, of which the first and the last are called the *extremes*, and the others the *means.*

The difference between the consecutive terms, or that quantity by which the numbers respectively increase upon each other, or decrease from each other, is called the *common difference.*

Thus, 3, 5, 7, 9, 11, &c., is an ascending arithmetical progression, and 11, 9, 7, 5, 3, is a descending arithmetical progression. In these progressions, in both instances, 11 and 3 are the *extremes*, of which 11 is the *greater extreme*, and 3 is the *less extreme*. The numbers between these, (9, 7, 5,) are the *means.*

In every arithmetical progression, the sum of the extremes is equal to the sum of any two means that are equally distant from the extremes; and is, therefore, equal to *twice* the middle term, when the series consists of an odd number of terms. Thus, in the foregoing series, $3 + 11 = 5 + 9 = 7 \times 2$.

The *greater extreme*, the *less extreme*, the *number of terms*, the

ARITHMETICAL PROGRESSION.

common difference, and the *sum of the terms,* are called the *five* properties of an arithmetical progression, of which, any *three* being given, the other two may be found.

Let s represent the sum of the terms.
" E "   the greater extreme.
" e "   the less extreme.
" d "   the common difference.
" n "   the number of terms.

The extremes of an arithmetical progression and the number of terms being given, to find the sum of the terms.

$$\frac{(E + e) \times n}{2} = \text{sum of the terms.}$$

EXAMPLE. — What is the sum of all the even numbers from 2 to 100, inclusive?

$$102 \times 50 \div 2 = 2550. \quad Ans.$$

EXAMPLE. — How many times does the hammer of a common clock strike in 12 hours?

$$(1 + 12) \times 12 \div 2 = 78 \text{ times.} \quad Ans.$$

$$\left(\frac{E-e}{d} + 1\right) \times \frac{E+e}{2} = \text{sum of the terms.}$$

$$(E \times 2 - \overline{n-1 \times d}) \times \tfrac{1}{2} n = \text{sum of the terms.}$$

$$(2e + \overline{n-1 \times d}) \times \tfrac{1}{2} n = \text{sum of the terms.}$$

The greater extreme, the common difference, and the number of terms of an arithmetical progression being given, to find the less extreme.

$$E - (d \times \overline{n-1}) = \text{less extreme.}$$

EXAMPLE. — A man travelled 18 days, and every day 3 miles farther than on the preceding; on the last day he travelled 56 miles; how many miles did he travel the first day?

$$56 - (\overline{18-1} \times 3) = 5 \text{ miles.} \quad Ans.$$

$$\frac{s}{n} - \left(\frac{\overline{n-1} \times d}{2}\right) = \text{less extreme.}$$

$$\frac{s}{n} \times 2 - E = \text{less extreme.}$$

$$\frac{\sqrt{(E \times 2 + d)^2 - s \times d \times 8} + d}{2} = \text{less extreme, when}$$

$\sqrt{(2E + d)^2 - 8sd}$ is equal to, or greater than d.

$$\frac{\sqrt{(2E + d)^2 - 8sd} \smallsmile d}{2} = \text{less extreme, when}$$

$\sqrt{(2E + d)^2 - 8sd}$ is less than d.

$$\frac{\sqrt{(2e \smallsmile d)^2 + 8sd} - d}{2} = \text{greater extreme}$$

$$d \times \overline{n-1} + e = \text{greater extreme.}$$

$$\frac{s}{n} + \frac{\overline{n-1} \times d}{2} = \text{greater extreme.}$$

$$2s \div n - e = \text{greater extreme.}$$

The extremes of an arithmetical progression and the common difference being given, to find the number of terms.

$$E - e \div d + 1 = \text{number of terms.}$$

EXAMPLE. — As a heavy body, falling freely through space, descends $16\frac{1}{12}$ feet in the first second of its descent, $48\frac{3}{12}$ feet in the next second, $80\frac{5}{12}$ in the third second, and so on; how many seconds had that body been falling, that descended $305\frac{7}{12}$ feet in the last second of its descent?

$$305\tfrac{7}{12} - 16\tfrac{1}{12} = 289\tfrac{1}{2} \div 32\tfrac{1}{6} = 9 + 1 = 10 \text{ seconds. } Ans.$$

$$\frac{\sqrt{(2e \smallsmile d)^2 + 8sd} - d}{2} - e \div d + 1 = \text{number of terms.}$$

$$2s \div E + \frac{\sqrt{(2E + d)^2 - 8sd} + d}{2} = \text{number of terms when}$$

$\sqrt{(2E + d)^2 - 8sd}$ is equal to, or greater than d.

$$2s \div E + \frac{\sqrt{(2E + d)^2 - 8sd} \smallsmile d}{2} = \text{number of terms when}$$

$\sqrt{(2E + d)^2 - 8sd}$ is less than d.

$$\frac{s \times 2}{E + e} = \text{number of terms.}$$

The extremes of an arithmetical progression, and the number of terms being given, to find the common difference.

$$\frac{E - e}{n - 1} = \text{common difference.}$$

EXAMPLE.—One of the extremes of an arithmetical progression is 28 and the other is 100, and there are 19 terms in the series; required the common difference.

$$100 \backsim 28 \div 1 \backsim 19 = 4. \quad Ans.$$

$$E - e \div \left(\frac{s \times 2}{E + e} - 1 \right) = \text{common difference.}$$

$$\frac{2s \div n - 2e}{n - 1} = \text{common difference.}$$

$$\frac{2E - (2s \div n)}{n - 1} = \text{common difference.}$$

EXAMPLE.—The less extreme of an arithmetical progression is 28, the sum of the terms 1216, and the number of terms 19; required the 7th term in the series, descending.

$$1216 \times 2 \div 19 = 128 = \text{sum of the extremes.}$$
$$128 - 28 = 100 = \text{greater extreme.}$$
$$100 - 28 = 72 = \text{difference of extremes.}$$
$$72 \div \overline{n - 1} \,(18) = 4 = \text{common difference.}$$
$$100 - (\overline{7 - 1} \times 4) = 76 = \text{7th term descending.} \quad Ans.$$

Required the 5th term from the less extreme, in an arithmetical progression, whose greatest extreme is 100, common difference 4, and number of terms 19.

$$100 - (\overline{19 - 5} \times 4) = 44. \quad Ans.$$

To find any assigned number of arithmetical means, between two given numbers or extremes.

RULE.—Subtract the less extreme from the greater, divide the remainder by one more than the number of means required, and the quotient will be the common difference between the extremes; which, added to the less extreme, gives the least mean, and, added to that, gives the next greater, and so on.

Or, $\overline{E - e} \div \overline{m + 1} = d$, E being the greater extreme, e the less extreme, m the number of means required, and d the common difference.

And $e + d$, $e + 2d$, $e + 3d$, &c.; or, $E - d$, $E - 2d$, $E - 3d$, &c., will give the means required.

Example.—Required to find 5 arithmetical means between the numbers 18 and 3.

$$18 - 3 = 15 \div 6 = 2\tfrac{1}{2}, \text{ and}$$
$$3 + 2\tfrac{1}{2} = 5\tfrac{1}{2} + 2\tfrac{1}{2} = 8 + 2\tfrac{1}{2} = 10\tfrac{1}{2} + 2\tfrac{1}{2} = 13 + 2\tfrac{1}{2} = 15\tfrac{1}{2}.$$

$5\tfrac{1}{2}, 8, 10\tfrac{1}{2}, 13, 15\tfrac{1}{2}$, therefore, are 5 arithmetical means, between the extremes, 3 and 18.

Note.—The arithmetical mean between any two numbers may be found by dividing the sum of those numbers by 2; thus, the arithmetical mean of 9 and 8 is $(9+8) \div 2 = 8\tfrac{1}{2}$.

GEOMETRICAL PROGRESSION.

A series of three or more numbers, increasing by a common multiplier, or decreasing by a common divisor, is called a *geometrical progression*. If the greater numbers of the progression are to the right, the progression is called an *ascending geometrical progression*, but, on the contrary, if they are to the left, it is called a *descending geometrical progression*. The number by which the progression is formed, that is, the common multiplier, or divisor, is called the *ratio*.

The numbers forming the series are called the *terms* of the progression, of which the first and the last are called the *extremes*, and the others the *means*. The greater of the extremes is called the *greater extreme*, and the less the *less extreme*.

Thus, 3, 6, 12, 24, 48, is an ascending geometrical progression, because 48 is as many times greater than 24, as 24 is greater than 12, &c.; and 250, 50, 10, 2, is a descending geometrical progression, because 2 is as many times less than 10, as 10 is less than 50, &c.

In the first mentioned series, (3, 6, 12, 24, 48,) 48 is the *greater extreme*, and 3 is the *less extreme;* the numbers 6, 12, 24 are the means in that progression.

So, too, of the progression 250, 50, 10, 2; 250 and 2 are the extremes, and 50 and 10 are the means.

In the first mentioned progression, 2 is the ratio, and in the last, or in the progression 2, 10, 50, 250, 5 is the ratio.

In a geometrical progression, the product of the two extremes is equal to the product of any two means that are equally distant from the extremes, and, also, equal to the square of the middle term, when the progression consists of an odd number of terms.

Thus, in the progression 2, 6, 18, 54, 162; $162 \times 2 = 54 \times 6 = 18 \times 18$.

When a geometrical progression has but 3 terms, either of the extremes is called a *third proportional* to the other two; and the middle term, consequently, is a *mean proportional* between them.

Thus, in the progression 48, 12, 3, 3 is a *third proportional* to 48 and 12, because 48 divided by the ratio = 12, and 12 divided by the ratio = 3; or 3 × ratio = 12, and 12 × ratio = 48 : 12 is the *mean proportional*, because 12 × 12 = 48 × 3.

Of the 5 properties of a geometrical progression, viz., the *greater extreme*, the *less extreme*, the *number of terms*, the *ratio*, and the *sum of the terms*, any three being given, the other two may be found.

Let s represent the sum of the terms.
" E " the greater extreme.
" e " the less extreme.
" r " the ratio.
" n " the number of terms.
" n when affixed as an index or exponent, represent that the term, number, or quantity, to which it is affixed, is to be raised to a power equal to the number of terms in the respective progression, &c.

Any three of the five parts of a geometrical progression being given, to find the remaining two parts.

$$\frac{E-e}{r-1} + E = \text{sum of the terms.}$$

$$\frac{E \times r - e}{r-1} = \text{sum of the terms.}$$

$$\frac{r^n \times e - e}{r-1} = \text{sum of the terms.}$$

$$\frac{E - (E \div r^{n-1})}{r-1} + E = \text{sum of the terms.}$$

$$\frac{E-e}{\sqrt[n-1]{(E \div e)} - 1} + E = \text{sum of the terms.}$$

EXAMPLE. — The greater extreme of a geometrical progression is 162, the less extreme is 2, and there are 5 terms in the progression; required the sum of the series.

$$\frac{162-2}{\sqrt[4]{(162 \div 2)} - 1} = 80 + 162 = 242. \quad Ans.$$

$$\frac{s \times \overline{r-1} + e}{r} = \text{greater extreme.}$$

$$\frac{r^n}{r} \times e = \text{greater extreme.}$$

$$r^{\overline{n-1}} \times e = \text{greater extreme.}$$

$$\frac{s \times r^{\overline{n-1}} \times \overline{r-1}}{r^n - 1} = \text{greater extreme.}$$

$$s - \overline{(s - E) \times r} = \text{less extreme.}$$

$$E \div r^{\overline{n-1}} = \text{less extreme.}$$

$$\frac{s \times r^{\overline{n-1}} \times \overline{r-1}}{r^n - 1} \div r^{n-1} = \text{less extreme.}$$

$$\frac{s - e}{s - E} = \text{ratio.}$$

$$\sqrt[n-1]{\frac{E}{e}} = \text{ratio.}$$

$\frac{s \times \overline{r-1}}{e} + 1 = r^n$; n, therefore, is equal to the number of times that r must be multiplied into itself to equal $\frac{s \times \overline{r-1}}{e} + 1$.

$$\frac{s \times \overline{r-1}}{s - \overline{(s-E) \times r}} + 1 = r^n.$$

EXAMPLE. — A farmer proposed to a drover that he would sell him 12 sheep and allow him to select them from his flock, provided the drover would pay 1 cent for the first selected, 3 cents for the second, 9 cents for the third, and so on; what sum of money would 12 sheep amount to, at that rate?

$$\frac{r^n \times e - e}{r - 1} = s, \text{ then}$$

$$\frac{3^{12} \times 1 - 1}{3 - 1} = \$2657.20, \text{ Ans.}$$

NOTE. — Ratio^4, cubed = ratio^{12}; ratio^6, squared = ratio^{12}, &c.

When it is required to find a high power of a ratio, it is convenient to proceed as follows, viz.: write down a few of the lower or leading powers of the ratio, successively as they arise, in a line, one after another, and place their respective indices over them; then

GEOMETRICAL PROGRESSION. 153

will the product of such of those powers as stand under such indices whose sum is equal to the index of the required power, equal the power required.

EXAMPLE. — Required the 11th power of 3.

$$\begin{array}{ccccc} 1 & 2 & 3 & 4 & 5 \\ 3 & 9 & 27 & 81 & 243 \end{array}$$

Here $5 + 4 + 2 = 11$, consequently,

$243 \times 81 \times 9 = $ 11th power of 3, or

$5 \times 2 + 1 = 11$, consequently,

$243 \times 243 \times 3 = $ 11th power of 3, or

$4 \times 2 + 3 = 11$, consequently,

$81 \times 81 \times 27 = $ 11th power of 3, or

$3 \times 3 + 2 = 11$, consequently,

$27^3 \times 9 = r^{11} = 177147.$ *Ans.*

To find any assigned number of geometrical means, between two given numbers or extremes.

RULE. — Divide the greater given number by the less, and from the quotient extract that root whose index is 1 more than the number of means required; that is, if 1 mean be required, extract the square root; if two, the cube root, &c., and the root will be the common ratio of all the terms; which, multiplied by the less given extreme, will give the least mean; and that, multiplied by the said root, will give the next greater mean, and so on, for all the means required. Or the greater extreme may be divided by the common ratio, for the greatest mean; that by the same ratio, for the next less, and so on.

EXAMPLE. — Required to find 5 geometrical means between the numbers 3 and 2187.

$2187 \div 3 = 729$, and $\sqrt[6]{729} = 3$, then —

$3 \times 3 = 9 \times 3 = 27 \times 3 = 81 \times 3 = 243 \times 3 = 729$, that is, the numbers 9, 27, 81, 243, 729 are the 5 geometrical means between 3 and 2187.

NOTE.— The geometrical mean between any two given numbers is equal to the square root of the product of those numbers. Thus the geometrical mean between 5 and 20, = $\sqrt{(5 \times 20)} = 10$.

ANNUITIES.

An annuity, strictly speaking and practically, is a certain sum of money by the year; payable, usually, either in a single payment yearly, or in half, half-yearly, quarter, quarter-yearly, &c., and for a succession of years, greater or less, or forever. *Pensions, awards, bequests,* and the like, that are made payable in fixed sums for a succession of payments, are commonly rated by the year, and denominated *annuities*.

A current annuity that has already commenced, or that is to commence after an interval of time not greater than that between the stipulated payments, is said to be in *possession*.

One that is to commence or cease on the occurrence of an indeterminate event, as upon the death of an individual, is a *reversionary, contingent,* or *life* annuity.

One that is to commence at a given period, and to continue for a given number of years or payments, is a *certain* annuity.

One that is to continue from a given time, forever, is a *perpetual annuity*, or a *perpetuity*.

Annuity payments do not exist fractionally: they mature, and exist only in that state, and are then due.

A current annuity commences with a payment, and terminates with a payment.

One current in the past is measured from a present included payment, closes with an included payment, and is said to be in *arrears* or *forborne*, from a supposed cancelled payment one regular interval or time beyond.

One current in the future is measured from the present to the first included payment of the series, and from thence is said to *continue* to the close; but if the interval from the present to the first included payment is equal to that between the successive payments, it is supposed to *continue* from the present.

Annuities in negotiation are adjusted, with regard to time, by *interest*, or *discount*, or both.

The TABLES applicable to compound interest and compound discount are applicable in adjusting annuities at compound rates.

To find the Amount of a Current Annuity in Arrears.

LEMMA. — The amount of an annuity that has been forborne for a given time is equal to the sum of the several payments that have become due in that time, plus the interest on each, from the time it became due, until the close of the time.

ANNUITIES. 155

Then the amount of an annuity of $100, payable in a single payment annually, but delayed of payment 4 years, allowing simple interest at 6 per cent. on the payments, is

$$100 \times 1.18 = 118$$
$$100 \times 1.12 = 112$$
$$100 \times 1.06 = 106$$
$$100 \times 1 = 100 = \$436.$$

And at 6 per cent. compound interest on the payments, it is

$$100 \times (1.06)^3 = 119.10$$
$$100 \times (1.06)^2 = 112.36$$
$$100 \times (1.06)^1 = 106.00$$
$$100 \times 1 = 100.00 = \$437.46.$$

At 6 per cent. simple interest, when payable in half, half-yearly, it is

$$50 \times 1.21 = 60.50$$
$$50 \times 1.18 = 59.00$$
$$50 \times 1.15 = 57.50$$
$$50 \times 1.12 = 56.00$$
$$50 \times 1.09 = 54.50$$
$$50 \times 1.06 = 53.00$$
$$50 \times 1.03 = 51.50$$
$$50 \times 1 = 50.00 = \$442.$$

And at 6 per cent. compound interest PER ANNUM, when payable in half-yearly instalments, it is

$$50 \times (1.06)^3 \times 1.03 = 61.34$$
$$50 \times (1.06)^3 = 59.55$$
$$50 \times (1.06)^2 \times 1.03 = 57.86$$
$$50 \times (1.06)^2 = 56.18$$
$$50 \times (1.06)^1 \times 1.03 = 54.59$$
$$50 \times 1.06 = 53.00$$
$$50 \times 1.03 = 51.50$$
$$50 \times 1. = 50.00 = \$444.02.$$

From the foregoing, we derive the following general RULES:—

Let $P =$ annuity or yearly sum,
$r =$ rate of interest per annum,
$a =$ rate of discount per annum,
n or $^n =$ nominal time of the annuity in full years,
$A = $ _amount_ for the full years,
$D = $ _present worth_ for the full years.

When the annuity is payable in a single payment yearly,

$$A = Pn\left(1 + \frac{r(n-1)}{2}\right), \text{ Simple Interest.}$$

$$A = P\frac{(1+r)^n - 1}{r}, \text{ Compound Interest.}$$

When payable in equal half-yearly instalments,

$$A = Pn\left(1 + \frac{r(n-1)}{2} + \frac{r}{4}\right), \text{ Simple Interest.}$$

$$A = P \times \frac{(1+r)^n - 1}{r} \times \left(1 + \frac{r}{4}\right), \text{ Compound Interest.}$$

When payable in equal third-yearly instalments,

$$A = Pn\left(1 + \frac{r(n-1)}{2} + \frac{r}{3}\right), \text{ Simple Interest.}$$

$$A = P\frac{(1+r)^n - 1}{r}\left(1 + \frac{r}{3}\right), \text{ Compound Interest.}$$

When payable in quarter-yearly instalments,

$$A = Pn\left(1 + \frac{r(n-1)}{2} + \frac{3r}{8}\right), \text{ Simple Interest.}$$

$$A = P\frac{(1+r)^n - 1}{r}\left(1 + \frac{3r}{8}\right), \text{ Compound Interest.}$$

When there are odd payments, to find the amount, S.

When 1 half-yearly, $S = A(1 + \frac{1}{2}r) + \frac{1}{2}P.$
1 third-yearly, $S = A(1 + \frac{2}{3}r) + \frac{1}{3}P.$
2 " $S = A(1 + \frac{2}{3}r) + \frac{1}{3}P(1 + \frac{1}{3}r) + \frac{1}{3}P$
$\quad = A(1 + \frac{2}{3}r) + P(6 + r) \div 9.$
1 quarter-yearly, $S = A(1 + \frac{1}{4}r) + \frac{1}{4}P.$
2 " . $S = A(1 + \frac{1}{2}r) + P(8 + r) \div 16.$
3 " $S = A(1 + \frac{3}{4}r) + P(3 + \frac{3}{4}r) \div 4.$

For any number of equal and regular payments at compound interest per interval between the payments, $S = P'\left(\frac{(1+r')^{n'} - 1}{r'}\right)$, and for any number of equal and regular payments at simple interest per interval between the payments, $S = P'n'\left(1 + \frac{r'(n'-1)}{2}\right)$; P' being a payment, n' or " the number of payments, and r' the rate of interest per interval between the payments. But this must not be confounded with compound interest *annually*, on payments occurring semi-annually, quarterly, &c.

EXAMPLE. — What is the amount of an annuity of $150, payable in half, half-yearly, but delayed of payment 2 years and 72 days, allowing compound interest per annum at 7 per cent. ?

$150 \times \frac{(1.07)^2 - 1}{.07} = \$310.50$, the amount for 2 years, if payable in yearly payments, and

ANNUITIES. 157

$310.50 \times \left(1 \cdot \frac{.07}{4}\right) = \$315.93$, the amount for 2 years, if payable in half-yearly payments, and

$315.93 \times \left(1 \cdot \frac{.07 \times 72}{365}\right) = \$320.29$, the amount for 2 years and 72 days, if payable in half-yearly payments. *Ans.*

EXAMPLE. — What is the amount of an allowance, pension, or award, of $100 a year, payable quarterly, but forborne $3\frac{1}{2}$ years, interest compound per annum at 6 per cent. ?

$100 \times \frac{(1.06)^3 - 1}{.06} \times \left(1 + \frac{.06 \times 3}{8}\right) = \$325.52$, the amount for 3 years, and

$325.52 (1 + .03) + 100 \times 8.06 \div 16 = \$385.66$. *Ans.*

EXAMPLE. — What is the amount of $100 a year, payable in quarterly payments, and in arrears 4 years, interest being compound per quarter-year, at 6 per cent. a year ?

$25 \left[\left(1 + \frac{.06}{4}\right)^{16} - 1\right] \times \frac{4}{.06}$. By tabular powers of $(1 + r)$, page 125, $= \$448.30$. *Ans.*

To find the Present Worth of an Annuity Current.

LEMMA. — The present worth of an annuity that is to *continue* for a given time is equal to that sum of money, which, if put at interest from the present time to the close of the payments, will amount to the amount of the payments at that time; and therefore, the times being full, is equal to the sum of the several payments, discounted, respectively, at the rate of interest for their respective times.

NOTE.—If the foregoing proposition is tenable, it follows, since simple interest is due and payable annually, that the true present worth of an annuity having more than one year to run cannot be found by simple interest and discount. By simple interest and discount, at 6 per cent., predicating the rule upon the foregoing lemma, the *amount* of $100, payable annually, and in arrears for 4 years, is $436; and the *present worth*, at 6 per cent., is

$$\frac{100}{1.24} + \frac{100}{1.18} + \frac{100}{1.12} + \frac{100}{1.06} = \$349.$$

But $349 at 6 per cent. interest for 4 years, with the payments of interest annually, will amount to $440.60; and at interest simply for 4 years it will amount to only $432.76.

Then the present worth of an annuity of $100, payable in a single payment yearly, and to continue 4 years, or to become due 1, 2, 3, and 4 years hence, interest and discount being compound per annum, and each at 6 per cent. =

$$\frac{P}{(1+r)^4}+\frac{P}{(1+r)^3}+\frac{P}{(1+r)^2}+\frac{P}{1+r}=\$346.51=$$

$100 \times (1.06)^3 = 119.10$
$100 \times (1.06)^2 = 112.36$
$100 \times (1.06)\ \ = 106.00$
$100 \times 1\ \ \ \ \ \ = 100.00 = 437.46 \div (1.06)^4 = \$346.51.$

And interest at 6 per cent. and discount at 10, both compound, it is

$100 \times (1.06)^3 = 119.10$
$100 \times (1.06)^2 = 112.36$
$100 \times\ 1.06\ \ = 106.00$
$100 \times 1\ \ \ \ \ \ = 100.00 = 437.46 \div (1.10)^4 = \$298.79.$

Therefore, when the annuity is payable in a single payment yearly from the present time,

$$D = P \frac{(1+r)^n-1}{r(1+a)^n} = \frac{A}{(1+r)^n} \text{ when } r \text{ and } a \text{ are equal.}$$

When payable in half-yearly payments,

$$D = P \times \frac{(1+r)^n-1}{r(1+a)^n} \times (1+\tfrac{1}{4}r).$$

When payable in third-yearly payments,

$$D = \frac{P \times [(1+r)^n - 1] \times (1+\tfrac{1}{3}r)}{r(1+a)^n}.$$

When payable in quarter-yearly payments,

$$D = \frac{P[(1+r)^n - 1](1+\tfrac{3}{8}r)}{r(1+a)^n}.$$

When there are odd payments, to find the present worth, S.

There being a half-yearly, $S = \frac{D}{1+\frac{1}{2}a} + \frac{\frac{1}{2}P}{1+\frac{1}{2}a}.$

" 1 third-yearly, $S = \frac{D}{1+\frac{1}{3}a} + \frac{\frac{1}{3}P}{1+\frac{1}{3}a}.$

" 2 " $S = \frac{D}{1+\frac{1}{3}a} + \frac{2P(1+\frac{1}{6}r)}{3(1+\frac{1}{3}a)}.$

" 1 quarter-yearly, $S = \frac{D}{1+\frac{1}{4}a} + \frac{\frac{1}{4}P}{1+\frac{1}{4}a}.$

" 2 " $S = \frac{D}{1+\frac{1}{4}a} + \frac{P(1+\frac{1}{8}r)}{2(1+\frac{1}{4}a)}.$

" 3 " $S = \frac{D}{1+\frac{1}{4}a} + \frac{\frac{3}{4}P(1+\frac{1}{4}r)}{1+\frac{1}{4}a}.$

For any number of equal payments, at equal intervals between the payments, $S = P' \times \frac{(1+r')^{n'}-1}{r'(1+a')^{n'}}$; P' being a payment, n' the

ANNUITIES.   159

number of payments, and r' and a' the rates per interval between the payments.

NOTE. — Since $\frac{(1+r)^a - 1}{r(1+a)^a}$ is the co-efficient of P, for its present worth, at compound interest and discount, for the time a, at the rates r, a, it follows that tables of co-efficients of P for its present worth, at given rates, for any number of years, may be easily made. Thus $(1.06^4 - 1) \div 1.06^4 \times .06 = 3.46511$, the co-efficient of an annuity, P, for 4 years' continuance, interest and discount being compound per annum, at 6 per cent.; and $(1.06^2 - 1) \div (1.06^2 \times .06) = 1.83339$, the co-efficient for 2 years, &c.

If the annuity is deferred, then the difference of two of these co-efficients (one of them that for the time deferred, and the other that for the sum of the time deferred and the time of the annuity) will be the co-efficient of P for its present worth. Thus $3.46511 - 1.83339 = 1.63172$, the co-efficient of an annuity, P, for its present worth, when it is to commence two years hence, and to continue 2 years, interest and discount being compound per annum, at 6 per cent. each; or $D = 1.63172$ P.

In like manner, tables of other co-efficients, such as the formulæ suggest, may be made that will greatly assist in calculating annuities.

EXAMPLE. — What is the present worth of an award of $500 a year, payable in half-yearly instalments, the 1st payment to mature 6 months hence, and the annuity to continue three years; interest and discount being 7 per cent., compounded yearly?

$$\frac{500 \times [(1.07)^3 - 1] \times \left(1.\frac{.07}{4}\right)}{.07 \times (1.07)^3} = \$1335.13. \quad Ans.$$

EXAMPLE. — What is the present worth of an annuity of $100, payable in half-yearly payments, and to continue $1\frac{1}{2}$ years; interest and discount being 6 per cent. per annum?

$$D = \frac{100 \times [1.06 - 1] \times 1.\frac{.06}{4}}{.06 \times 1.06} = 95.755, \text{ and}$$

$$\frac{95.755}{1.03} + \frac{50}{1.03} = \$141.51. \quad Ans.$$

EXAMPLE. — What is the present worth of an annuity of $500, payable in semi-annual instalments, and to continue $10\frac{1}{2}$ years, interest and discount being compound per annum, the former at 6 per cent., and the latter at 8?

$$\frac{500[(1.06)^{10} - 1]\left(1.\frac{.06}{4}\right)}{.06(1.08)^{10}\left(1.\frac{.08}{2}\right)} + \frac{500}{2\left(1.\frac{.08}{2}\right)} =$$

$$\frac{A}{1.08^{10} \times 1.04} + \frac{250}{1.04} = \quad Ans.$$

By tabular powers of $1 + r$, page 125:—

$$\frac{500 \times .79085}{.06 \times 2.15892} = \$3052.64,$$ the present worth for 10 years' continuance, if payable in yearly payments, and

$$3052.64 \times 1.015 = \$3098.43,$$

the present worth for 10 years' continuance, if payable in half-yearly payments, and

$$3098.43 \div 1.04 + 500 \div 2 \times 1.04 = \$3219.64. \quad Ans.$$

When the interval of time from the present to the 1st payment is *shorter* than that between the consecutive payments, and the annuity is payable in a single payment yearly,

$$A = \frac{P\left[(1+r)^n - 1\right]\left(1 + \frac{dr}{365}\right)}{r}, \text{ and}$$

$$D = \frac{A}{(1+a)^{(n-1)}\left(1 + \frac{a(365-d)}{365}\right)} = \frac{P\left[(1+r)^n - 1\right]\left(1 + \frac{dr}{365}\right)}{r(1+a)^{(n-1)}\left(1 + \frac{a(365-d)}{365}\right)},$$

d being the time in days from the present to the 1st payment.

So, if the annuity is payable in half-yearly, third-yearly, or quarter-yearly instalments, multiply by $1 + \frac{1}{4}r$, $1 + \frac{1}{6}r$, or $1 + \frac{3}{8}r$, as before directed; and if there are odd payments proceed for the present worth, S, as already directed.

EXAMPLE. — Required the present worth of an annuity of $100, payable yearly, to commence 4 months hence, and to continue 4 years; interest and discount being 6 per cent. *annually*.

$$\frac{100 \times (1.06^4 - 1) \times \left(1.\frac{4 \times .06}{12}\right)}{.06 \times 1.06^3 \times \left(1.\frac{.06 \times (12-4)}{12}\right)} = \$360,24. \quad Ans.$$

To find the Present Worth of a Deferred Current Annuity, or of an Annuity in Reversion.

When the annuity is payable in a single payment yearly, and the deferred time embraces full years only,

$$D = P\frac{(1+r)^n - 1}{r(1+a)^{(n+n')}}, \text{ n' being the deferred time.}$$

If it is payable in half-yearly, third-yearly, or quarter-yearly instalments, multiply by $1 + \frac{1}{4}r$, $1 + \frac{1}{6}r$, or $1 + \frac{3}{8}r$, as already directed; and, if there are odd payments, find the present worth, S, as already directed.

EXAMPLE. — What is the present worth of an annuity of $150, payable yearly, to commence 2 years hence, and to continue 4 years; interest and discount being compound per annum, at 6 per cent. ?

$$150 \times (1.06^4 - 1) \div .06 \times 1.06^6 = \$462.59. \quad Ans.$$

EXAMPLE. — Required the present worth of an annuity of $500, payable in semi-annual instalments, to commence 2½ years hence, and to continue 6 years; allowing compound interest and discount annually at 7 per cent.

$$\frac{500 \times (1.07^6 - 1) \times 1.\frac{.07}{4}}{.07 \times 1.07^8 \times 1.\frac{.07}{2}} = \$2046.44. \quad Ans.$$

EXAMPLE. — Required the present worth of an allowance, pension, or award of $125 a year, payable in half every half-year, to commence 7 months 24 days hence, and to continue 6½ years; interest and discount being compound per annum at 5 per cent.

$$\frac{125 \times (1.05^6 - 1) \times 1.0125}{.05 \times 1.05^6 \times 1.03247 \times 1.025} + \frac{125}{2 \times 1.025} = \$668. \quad Ans.$$

Or $\dfrac{125 \times [(1.05)^6 - 1]}{.05 \times (1.05)^6} = \$634.47$, the present worth for 6 years' continuance, if payable in yearly instalments; and

$$634.47 \times 1.\frac{.05}{4} = \$642.40,$$

the present worth for 6 years' continuance, if payable in half-yearly instalments; and

$$642.40 \div \left(1 + \frac{.05 \times 237}{365}\right) = 622.20,$$

the present worth for 6 years' continuance, if payable in half-yearly instalments, and to commence 7 months, 24 days hence; and

$$622.20 \div \left(1 + \frac{.05}{2}\right) + \frac{125}{2 \times 1.025} = \$668. \quad Ans.$$

To find the Present Worth of a Perpetuity.

LEMMA. — The present worth of an annuity to commence one year hence, and to continue forever, is expressed by that sum of money whose interest for 1 year is equal to the *amount* of the annuity for 1 year; and so, *pro rata*, for perpetuities otherwise regularly affected.

Then when the annuity is to commence 1 year hence, and is payable in a single payment yearly . . . $D = P \div r$.

Payable in half-yearly instalments . . $D = \dfrac{P(1 + \frac{1}{4}r)}{r}$.

Payable in third-yearly instalments . . $D = \dfrac{P(1 + \frac{1}{3}r)}{r}$.

Payable in quarter-yearly instalments . $D = \dfrac{P(1 + \frac{3}{8}r)}{r}$.

EXAMPLE. — What is the present worth of a perpetuity of $150 a year, payable in a single payment yearly from the present time; interest at 6 per cent?

$$150 \div .06 = \$2500. \quad Ans.$$

EXAMPLE. — What is the present worth of a perpetuity of $150 a year, payable in semi-annual instalments, and to commence 4 months hence; interest 7 per cent?

$$\dfrac{P(1 + \frac{1}{4}r)}{r} + P \cdot \dfrac{.07(12 - 4)}{12} = \$2187.36. \quad Ans.$$

EXAMPLE. — Required the present worth of a perpetuity of $400 a year, payable in quarterly payments, and to commence 6 years hence; interest and discount being 5 per cent., compound per year.

$$D = \dfrac{P(1 + \frac{3}{8}r)}{r(1 + a)^a} = \dfrac{400 \times 1.\frac{3 \times .05}{8}}{.05 \times 1.05^6} = \$6081.65. \quad Ans.$$

The Amount, Time, and Rate given, to find the Annuity.

When payable in a single payment yearly from the present time,

$P = \dfrac{Ar}{(1 + r)^n - 1}$; half-yearly, $P = \dfrac{Ar}{[(1 + r)^n - 1](1 + \frac{1}{4}r)}$;

third-yearly, $P = \dfrac{Ar}{(1 + \frac{1}{3}r)[(1 + r)^n - 1]}$; quarterly,

$$P = \dfrac{Ar}{(1 + \frac{3}{8}r)[(1 + r)^n - 1]};$$

and so, *pro rata*, for other fractional units of the integral unit.

Therefore $(1+r)^a - 1 = \frac{Ar}{P}$, or $\frac{Ar}{P(1+\frac{1}{2}r)}$, or $\frac{Ar}{P(1+\frac{1}{3}r)}$ or $\frac{Ar}{P(+\frac{3}{8}r)}$, &c.

EXAMPLE. — What annuity, payable in quarterly payments from the present time, will amount to $3000 in 12 years; interest, being compound per annum, at 8 per cent. ?

$$3000 \times .08 \div \left[(1.08^{12}-1)\times 1.\tfrac{3\times .08}{8}\right] = \$153.48. \text{ Ans.}$$

EXAMPLE. — What length of time must a current annuity of $400, payable in quarterly payments, remain unpaid, that it may amount to $2500; interest being 7 per cent. yearly ?

$$\frac{2500\times.07}{400\times 1.\frac{3\times .07}{8}} = .4263094 = 5+ \text{ years, and 5 years by table of}$$

$(1+r)^n - 1 = .402552$: therefore $\left(\frac{.4263094}{.402552} - 1\right)\frac{365}{.07} = 308$ days, 5 years, 308 days. *Ans.*

The Present Worth, Time, and Rate given, to find the Annuity.

When payable in a single payment yearly from the present time,

$P = \frac{Dr(1+r)^n}{(1+r)^n - 1}$; half-yearly, $P = \frac{Dr(1+r)^n}{[(1+r)^n - 1](1+\frac{1}{2}r)}$; third-yearly, $P = \frac{Dr(1+r)^n}{[(1+r)^n - 1](1+\frac{1}{3}r)}$; quarter-yearly, $P = \frac{Dr(1+r)^n}{[(1+r)^n - 1](1+\frac{3}{8}r)}$, &c. Therefore, $(1+r)^n; = \frac{P}{P-Dr} =$

$$\frac{P(1+\frac{1}{2}r)}{P(1+\frac{1}{2}r) - Dr} = \frac{P(1+\frac{3}{8}r)}{P(1+\frac{3}{8}r) - Dr} \text{, \&c.}$$

EXAMPLE. — What annuity, payable in half-yearly instalments, and to continue 3 years, is at present worth $1335.13; discount and interest being compound per year, at 7 per cent ?

$$\frac{1335.13\times.07\times 1.07^3}{(1.07^3-1)\times 1.\frac{.07}{4}} = \$500. \text{ Ans.}$$

OF INSTALMENTS GENERALLY.

Any certain sum of money to be paid on a debt periodically until the debt is paid is called an instalment; and a debt so made payable is said to be payable by instalments.

Let $D =$ principal or debt to be paid,
$n =$ number of years in which the debt is to be paid,
$r =$ rate of interest per annum,
$p =$ instalment or periodical payment.

When the instalments are payable yearly, and the debt is at interest,

$$p = \frac{Dr(1+r)^n}{(1+r)^n - 1}; (1+r)^n = \frac{p}{p - Dr}; D = \frac{p[(1+r)^n - 1]}{r(1+r)^n}.$$

When payable half-yearly,

$$p = \frac{Dr(1+r)^n}{2[(1+r)^n - 1](1 + \tfrac{1}{4}r)};$$

$$(1+r)^n = \frac{p(1 + \tfrac{1}{4}r) + p}{[p(1+\tfrac{1}{4}r)+p] - Dr}; D = \frac{2p[(1+r)^n - 1](1 + \tfrac{1}{4}r)}{r(1+r)^n}.$$

When the debt is not on interest, and the instalments are payable yearly,

$$p = \frac{Dr}{(1+r)^n - 1}; (1+r)^n = \frac{Dr + p}{p}; D = \frac{p(1+r)^n - p}{r}.$$

EXAMPLE. — What yearly instalment will pay a debt of $4000 in 4 years, the debt being on interest the while, at 6 per cent. annually?

$4000 \times .06 \times 1.06^4 \div (1.06^4 - 1) = \$1154.37$. *Ans.*

EXAMPLE. — What semi-annual instalment will pay a debt of $4500 in 3 years, the debt bearing interest at 7 per cent. yearly?

$$\frac{4500 \times .07 \times 1.07^3}{2 \times (1.07^3 - 1) \times 1.0175} = \$842.62. \ \ Ans.$$

ANNUITIES.

When a debt has been diminished at regular intervals by the payment of a constant sum, to find the remaining debt at the close of the last payment.

When the debt is on interest, and the payments have been made yearly from the date of the debt,

$$d = \frac{p - (p - Dr)(1 + r)^n}{r}; \quad (1 + r)^n = \frac{p - dr}{p - Dr}$$

$$p = \frac{Dr(1 + r)^n - dr}{(1 + r)^n - 1}.$$

When the payments have been made half-yearly,

$$d = p + p(1 + \tfrac{1}{2}r) - (1 + r)^n [p + p(1 + \tfrac{1}{2}r) - Dr] \div r, \&c.$$

EXAMPLE. — On a debt of $1000, drawing interest the while at 8 per cent. a year, there has been paid yearly, from the date of the debt, $200 for 6 years: required the unpaid debt at the close of the last payment.

[200 — 1.08⁶(200 — .08 × 1000)] ÷ .08 = $119.69. *Ans.*

EXAMPLE. — On a note of hand for $1000, and interest from date, at 8 per cent. annually, the following payments have been made; viz., $100 at the close of every half-year from the date of the note, for 6 years. How much remained unpaid at the close of the last payment?

[204 — 1.08⁶(204 — .08 × 1000)] ÷ .08 = $90.34. *Ans.*

PERMUTATION.

PERMUTATION, in the mathematics, has reference to the greatest number of unlike relative positions, that a given number of things, either wholly unlike, or unlike only in part, may be placed in. It considers the number of changes, therefore, that may be made, in the arrangement of the things, under different given circumstances.

To find the number of changes that can be made in the order of arrangement of a given number of things, when the things are all different.

RULE. — Find the product of the natural series of numbers, from 1 up to the given number of things, inclusive; and that product will be the number of changes or permutations that may be made.

EXAMPLE. — In how many different relative positions may 12 persons be seated at a table?

$1 \times 2 \times 3 \times 4 \times 5 \times 6 \times 7 \times 8 \times 9 \times 10 \times 11 \times 12 = 479,001,600.$ *Ans.*

To find the number of changes that can be made in the order of arrangement of a given number of things, when that number is composed of several different things, and of several which are alike.

RULE. — Find the number of changes that could be made if the things were all unlike, as in first example. Then find the number of changes that could be made with the several things of each kind, if they were unlike. Lastly, divide the number first found by the product of the numbers last found, and the quotient will be the number of permutations or changes that the collection admits of.

EXAMPLE. — Required the number of permutations that can be made with the letters $a, bb, ccc, dddd, = 10$ letters.

$$\frac{1 \times 2 \times 3 \times 4 \times 5 \times 6 \times 7 \times 8 \times 9 \times 10 = 3628800}{1 \times 2 \times 6 \times 24 = 288} = 12,600. \text{ \textit{Ans.}}$$

To find the number of permutations that can be made with a given number of different things, by taking an assigned number of them at a time.

RULE. — Take a series of numbers beginning with the number of things given, and decreasing by 1 continually, until the number of terms is equal to the number of things that are to be taken at a time; then will the product of the series be the number of changes that may be made.

EXAMPLE. — What number of changes can be made with the numbers 1, 2, 3, 4, 5, 6, taking three of them at a time?
$6 - 1 = 5, 5 - 1 = 4$, then $6 \times 5 \times 4 = 120$. *Ans.*

What number, by taking 4 of them at a time?
$6 \times 5 \times 4 \times 3 = 360$. *Ans.*

EXAMPLE. — Arrange the three letters a, b, c, into the greatest number of permutations possible.

abc, acb, bac, bca, cab, cba, $= 6$ permutations. *Ans.*

EXAMPLE. — Arrange the four letters a, b, a, b, into the greatest number of permutations possible.

abab, aabb, abba, bbaa, baba, baab, $= 6$ permutations. *Ans*

COMBINATION.

COMBINATION, in the mathematics, has reference to the number of unlike groups, which may be formed from a given number of different things, by taking any assigned number of them, less than the whole at a time. It does not regard the relative positions of the things, one with another, in any of the collections or groups. But it exacts that each group, in all instances, shall have the assigned number of members in it, and that, in every group, in every instance, there shall be a like number of members. It exacts, therefore, that no two groups shall be composed of precisely the same members.

To find the number of combinations that can be made from a given number of different things, by taking any given number of them at a time.

RULE. — Take a series of numbers beginning with that which is equal to the number of things from which the combinations are to be made, and decreasing by 1, continually, until the number of terms is equal to the number of things that are to be taken at a time, and find the product of those numbers or terms. Then take the natural series, 1, 2, 3, 4, &c., up to the number of things that are to be taken at a time, and find the product of that series. Lastly, divide the product first found by the product last found, and the quotient will express the number of combinations that can be made.

EXAMPLE. — What number of combinations can be made from 8 different things, by taking 4 of them at a time?

$$\frac{8 \times 7 \times 6 \times 5}{1 \times 2 \times 3 \times 4} = \frac{1680}{24} = 70. \ Ans.$$

What number, by taking 5 of them at a time?

$$\frac{8 \times 7 \times 6 \times 5 \times 4}{1 \times 2 \times 3 \times 4 \times 5} = \frac{6720}{120} = 56. \ \textit{Ans.}$$

What number, by taking 3 of them at a time?

$$\frac{8 \times 7 \times 6}{1 \times 2 \times 3} = \frac{336}{6} = 56. \ \textit{Ans.}$$

EXAMPLE. — What number of combinations can be made from 5 different things, by taking three of them at a time?

$$\frac{5 \times 4 \times 3}{1 \times 2 \times 3} = \frac{60}{6} = 10. \ \textit{Ans.}$$

What number, by taking 2 of them at a time?

$$\frac{5 \times 4}{1 \times 2} = \frac{20}{2} = 10. \ \textit{Ans.}$$

EXAMPLE. — Form 5 letters, a, b, c, d, e, into 10 combinations of 2 letters each; that is, into 10 unlike groups of two letters each.

ab, ac, ad, ae, bc, bd, be, cd, ce, de. Ans.

Form them into the greatest number of combinations possible, in collections of three each.

abc, abd, abe, acd, ace, ade, bcd, bce, bde, cde. Ans.

FOREIGN MONEYS OF ACCOUNT:

THEIR DENOMINATIONS, RELATIVE VALUES, AND VALUES IN FEDERAL MONEY.

THE value in Federal money, affixed to any particular denomination of a Foreign money of account, in the following TABLES, is the intrinsic par thereof, as near as practicable. It is based upon the standard weight and purity of the coins coined especially to represent that denomination, compared with the standard weight and purity of the coins of the United States that represent the *dollar*, and is the United States Customs value of that denomination for computing duties. The denomination itself, to which the Federal value is immediately affixed, is usually the *unit* or ultimate money of account of the country especially referred to. It is a money of account in that country always; but not always in that country the name of a national coin. It is not always, even, represented by a single national coin. Thus, in Great Britain, until comparatively recently, there was no single British coin of the value of one *pound*. That denomination is now represented by a single gold coin, called a *sovereign*.

Foreign.	U. States.

ALGIERS. — *Algiers, Bona,* &c. : 100 centimes = 1 Livre, - - - - - - = $0.187
  29 aspers = 1 tomin, 8 tomins = 1 pataka-chicas,
  3 patakas-chicas = 1 Piastre, - - = 0.21
ARABIA. — *Mocha, Jidda*: 80 caveers = 1 Piastre, - = 0.823
  1 wakega *for gold and silver* = 480 troy grains.
AUSTRIA. — 100 centesimi = 1 Lira Austriache, - = 0.1622
  Vienna, Trieste, Prague (Commercial): 4 pfennige
  = 1 kreuzer, 60 k. = 1 Gulden *or* Florin Austriache = $\frac{1}{20}$ of a Cologne mark of fine silver =
  $180\frac{35}{100}$ troy grains, - - - - = 0.4858
  6 hellers = 1 groschen, 80 g. = 1 Florin, - = 0.4858
  1½ groschen = 1 kreuzer.
  1¼ florins = 1 current Thaler.
  2 florins = 1 Specie Thaler.
  1 florin Austriache = $5\frac{5}{17}$ lire de Trieste = $5\frac{7}{17}$ lire di piazza.
  1 mark Austrian = 4332¼ troy grains.

FOREIGN MONEYS OF ACCOUNT.

Foreign.	U. States.
AZORE ISLANDS. — *Corvo, Fayal, Flores, Graciosa, Pico, Terceira, St. Michael, St. Mary's:* 1000 reás = 1 Milrea,	= $0.83¼
BALEARIC ISLANDS. — Majorca I. — *Palma:* Minorca I. — *Port Mahon:* same as Cadiz, Spain.	
BELGIUM. — *Antwerp, Brussels, Liege, Mechlin, Ghent,* &c. :	
10 centimes = 1 decime, 10 d. = 1 Franc,	= 0.187
24 mitres *or* 8 Brabant penning *or* 6 liarde *or* 2 groote = 1 stuiver, 6 stuivers = 1 schelling, 20 s. = 1 Pond Flemish,	= 2.377
6 gulden *or* 2½ daalder = 1 Pond Flemish.	'-
BERMUDAS I. — 4 farthings = 1 penny, 12 pence = 1 shilling, 20 shillings = 1 Pound,	= 3.00
BOURBON ISLANDS. — 100 centimes = 1 Franc,	= 0.187
BRAZIL. — *Aracati, Bahia, Maranham, Para, Pernambuco, Portalegra, Rio Janeiro,* &c. : 1000 reás = 1 Milrea,	= 1.042
BONAIRE I. — 48 stuivers *or* 8 reáls = 1 Piastre,	= 0.73

Central and South America.

Balize, Campeche, Guatimala, Honduras, Laguna, Leon, Nicaragua, San Juan, San Salvador, Sisal, &c. :
Buenos Ayres, Callao, Carthagena, Coquimbo, Guayaquil, Laguayra, Lima, Maracaybo, Montevideo, Rio Hacha, Truxillo, Valparaiso, &c. :

2 cuartilli = 1 medio, 2 medios = 1 Reál,	= 0.12¼
8 reále = 1 Peso,	= 1.00
Pound of Honduras,	= 3.00
Berbice, Demerara, Essequibo, Surinam :	
8 duyt = 1 stuiver, 20 s. = 1 Florin *or* Gulden,	= 0.33¼
Cayenne : 100 centimes = 1 Franc *or.* Livre,	= 0.187
CANARY ISLANDS. — *Grand Canary, Teneriffe, Palma,* &c. :	
34 maravedi = 1 Reál (*current*),	= 0.074
CANDIA I. — 80 aspers *or* 33 medini = 1 Piastre,	= 0.048
CAPE VERD I. — 1000 reás = 1 Milrea,	=
CHILI. — *Coquimbo, San Carlos, Santiago, Valdivia, Valparaiso,* &c. :	
4 cuartilli = 1 medio, 2 m. = 1 reál, 8 r. = 1 Peso,	= 1.00
CHINA. — *Amoy, Canton, Macao, Nankin, Pekin, Shanghae,* &c. :	
10 cash = 1 candarine, 10 candarines = 1 mace, 10 mace = 1 Tael,	= 1.48
Hong Kong I. — Same as Great Britain.	

Foreign.	*U. States.*
CORSICA I. — 100 centimes = 1 Franc, - - =	$0.187
CYPRESS I. — Same as Constantinople (*Turkey*).	
DENMARK. — *Copenhagen*, &c. : 12 pfennige = 1 skilling, 16 s. = 1 mark, 6 m. = 1 Rykshaukdaler = $\frac{2}{37}$ of a Cologne marc, or 194.98 troy grains of pure silver, - - - - - =	0.5252
2 Rykshankdalers = 1 Speciesdaler, - - =	1.05
EGYPT. — *Alexandria* and the *Delta:* 8 borbi *or* 6 fiorli *or* 3 aspers = 1 medimne, 40 m. = 1 Piastre, =	0.048
20 piastres = 1 reál (*a gold coin*), - - - =	0.968
Cairo: 80 aspers *or* 33 medimni = 1 Piastre, - =	0.048
FRANCE. — Standard for gold and silver coins = $\frac{9}{10}$ fine, each : Relative values — gold to silver as 15 to 1.	
10 centimes = 1 decime, 10 d. = 1 Franc = $\frac{9}{2000}$ kilogrammes of fine silver = 69.449 troy grains, =	0.18706
12 deniers = 1 sou, 20 s. = 1 Livre tournois (*old*) = $\frac{80}{81}$ franc.	

GERMANY. — The mark of Cologne, divided into 8 unze = 64 quent = 256 pfennig = 512 heller = 4352 eschen = 65536 richtpfennig, is employed at the mints, for weighing gold and silver coins, throughout Germany : this mark = 3607¼ troy grains.

The standard for the Ducat, at present, throughout Germany, is $\frac{71}{72}$ or 23⅔ carats fine, its weight, $\frac{1}{67}$ of a Cologne mark. This piece, therefore, should weigh 53.84 troy grains, and contain 53.09 troy grains of fine gold = $2.286.

The standard for the Pistole or Zehn Gulden piece, of the Convention of 1753, is $\frac{87}{96}$ or 21¾ carats fine ; its weight, $\frac{1}{35}$ of a Cologne mark. This piece, therefore, should weigh 103.06 troy grains, and contain 93.4 troy grains of fine gold = $4.022.

The standard for the Specie Thaler, or Rixdollar affective, of the Convention of 1753, is ⅚ or 13 loth 6 gran fine, its weight $\frac{3}{25}$ Cologne mark = $0.9716 ; and the species Florin is ½ thereof = $0.4858.

The standard value of the Current Thaler or Rixdollar of account, established about 1775, is ¾ Rixdollar affective = 1½ specie Florins = $0.7287.

The Thaler of the Convention of 1838 contains $\frac{1}{14}$ Cologne mark of fine silver = $0.694 ; and the Gulden or Florin of that Convention = $\frac{4}{7}$ thereof = $0.3966 : 1¾ gulden = 1 Thaler.

The standard of purity for the coins established by the Convention of 1838, are : Two Thaler piece = $\frac{9}{10}$; thaler, ⅓ thaler, ⅙ thaler, each, = ¾ ; florin, ½ florin, each = $\frac{9}{10}$, less fractions = ¼.

Three Thalers of the Convention of 1838, by the estimate of relative values then established (gold to silver as 14.56 to 1), = 1 Ducat; while five of the current Rixdollars, established in 1775, by the then existing estimate of relative values (gold to silver as 14.483 to 1), = 1 Pistole, or 10 guilder piece, above referred to.

Most of the moneys of account throughout Germany, including Austria and Prussia, are based upon some portion of the foregoing.

Foreign.	U. States.
Bremen: 5 schwaren = 1 groot, 72 g. or 2 florins = 1 Rixdollar = ⅓ standard Carl d'or,	$0.78¾
Frankfort: 4 pfennig = 1 kreuzer, 60 k. = 1 Gulden = 1/10 Convention pistole,	0.4022
90 kreuzer = 1 Rixdollar of account,	0.6033
1½ albus = 1 kraiser-groschen, 30 k. = 1 Rixdollar,	0.6033
4 kreuzer = 1 batzen, 22½ b. = 1 Rixdollar,	0.6033
120 kreuzer *or* 2 gulden *or* 1½ Rixdollar = 1 Species Rixdollar;	0.8044
Hamburg, Lubec: 12 pfennig = 1 schilling, 16 schillinge = 1 mark.	
1 mark current = $0.28. 1 mark banco,	0.35
3 marks banco = 1 Specie Rixdollar,	1.05
BADEN. — *Carlsruhe, Heidelberg,* &c.: Same as BAVARIA.	
BAVARIA. — *Augsburg, Bamberg, Bayreuth, Munich, Nuremberg, Ratisbon, Wurtzburg,* &c.: 4 denari = 1 kreuzer, 60 k. = 1 Florin,	0.4858
2 kreuzer = 1 albus, 2 a. = 1 batzen, 15 batzens = 1 Florin.	
1½ florins = 1 Rixdollar current,	0.7287
2 florins = 1 Specie Rixdollar.	
HANOVER. — *Emden, Gottengen, Hanover, Osnaburg,* &c.: 12 pfennig = 1 gute groschen, 24 g. = 1 Rixdollar,	0.7287
1½ Rixdollar = 1 Specie Rixdollar.	
60 kreuzer = 1 gulden, 1¾ g. = 1 Thaler,	0.694
HESSE. — 12 hellers *or* 9 pfennig = 1 albus, 32 albus = 1 Rixdollar,	0.7287
24 marien groschen = 1 reichsflorin, 1½ r. = 1 Rixdollar.	
60 kreuzer = 1 gulden, 1¾ g. = 1 Thaler,	0.694
HOLSTEIN, *Altona, Kiel,* &c.: 12 pfennig = 1 schilling lubs, 16 schillinge (*of Lubec*) = 1 mark,	0.35
3 marks = 1 Speciesdaler *of Denmark.*	
MECKLENBURG. — *Rostock, Wismar,* &c.: Same as HANOVER.	
OLDENBURG. —Same as *Bremen;* also, same as *Hamburg.*	

| Foreign. | U. States. |

SAXONY. — *Dresden, Leipsic,* &c.: 12 pfennig = 1 gute groschen, 24 g. = 1 Species Thaler, - - = $0.7287
16 gute groschen = 1 reichflorin, 2 r. = 1 Specie Rixdollar.

SAXE. — *Gotha, Weimar:* Same as HANOVER.

SAXE *generally* and NASSAU: 4 hellers = 1 kreuzer, 60 kreuzers = 1 Gulden, - - - - = 0.4022
1½ gulden = 1 Rixdollar current.

WURTEMBURG. — *Halle, Stuttgard, Ulm,* &c.: Same as Saxe and Nassau.

GREAT BRITAIN. — *Sterling money:* Standard for silver coins = $\frac{37}{40}$ fine; for gold coins = $\frac{11}{12}$ fine. Relative values, gold to silver as 14.288 to 1.
4 farthings = 1 penny, 12 p. = 1 shilling, 20 s. = 1 Pound = $4.866.* U. S. Customs value, - = 4.84

GREECE. — 12 denari = 1 soldo, 20 s. = 1 Lira or Drachma, - - - - - = 0.163

HOLLAND. — Standard for silver coins = $\frac{945}{1000}$ fine. Standard for gold coins — Gouden Willem (10 *florins*) fractions and multiples = $\frac{900}{1000}$ fine — Ducat and multiples = $\frac{983}{1000}$ fine; weight of Ducat = 3.494 grammes = $2.2827. Relative values — gold to silver as 15.604 to 1.
100 centimes = 1 Florin *or* Gulden = 9.45 grammes of fine silver = 145.843 troy grains, - - = 0.3928
2½ florins = 1 Ryksdaalder, - - - = 0.982
10 florins in gold = 6.056 grammes fine gold = $4.0251. Custom-House value of Florin = $0.40.

India and Malaysia or East Indies.

BRITISH POSSESSIONS. — Standard of purity for gold and silver coins = $\frac{11}{12}$ fine, each.

HINDOSTAN. — *Bombay, Surat, Tatta:* 100 reas = 1 quarter, 4 q. = 1 Rupee = 165 troy grains of fine silver, - - - - - - = 0.4444

---

* For all time since 1816, the government of Great Britain has estimated gold and silver as 14.2879 to 1. The *pound sterling* of mint silver weighs 1745.454 grains, and contains 1614.545 grains of fine silver. The value of the *pound sterling* of silver, therefore, rated by the United States standard of 371¼ grains of fine silver to the *dollar*, is $4.349. The value of the silver shilling, of full weight, is $0.2174 The *pound sterling* of mint gold weighs 123.274 grains, and contains 113.001 grains of fine gold. The value of the *pound sterling* of gold, rated by the United States standard of 23.22 grains of fine gold to the *dollar*, is $4.866. At the *old* intrinsic par between the two currencies, viz., 4 *shillings* and 6 *pence sterling* to the *dollar*, or $4.44⁴⁄₉ to the £, the *par* of exchange is 9¼ per cent. Gold is the money standard in Great Britain, and the silver coins of that country are not legal tender at home in sums exceeding £2.

FOREIGN MONEYS OF ACCOUNT.

Foreign.	U. States.

Calcutta: 12 pice = 1 anna, 16 annas = 1 Rupee, = $0.4444
    1 Arcot rupee = 1 Sicca rupee = 0.44\frac{132}{187}$.
    1 Mohur or Gold Rupee = 165 troy grains fine gold
    = $7.1033.
    A Lac of rupees = 100.000 rupees.
    A Crore of rupees = 100 lacs of rupees.
MADRAS. — 12 pice = 1 anna, 16 annas = 1 Rupee, = 0.4444
    80 cash = 1 fanam, 42 fanams = 1 Pagoda, - = 1.851
    3½ old Sicca rupees = 1 Pagoda.
    1 Sicca rupee less 16 per cent. = 1 current rupee.
Cochin: 12 pice = 1 anna, 16 annas = 1 Rupee, = 0.4444
    4 fanams = 1 schilling, 5 schillings = 1 Rupee.
Goa: 18 budgerooks = 1 vintim, 4 v. = 1 tanga, 5 t.
    = 1 Pardo, - - - - - = 0.96
Pondicherry: 60 cash = 1 fanam, 24 f. = 1 Pagoda, = 1.84
BANCA I. — Same as BATAVIA (Java I.)
BORNEO I. — Borneo, Pentaniah, Sambos, &c.: Same as
    BATAVIA (Java I.)
CELEBES I. — Macassar: Same as BATAVIA (Java I.)
CEYLON I. — Colombo: 4 pice = 1 fanam, 12 fanams =
    1 Ryksdaalder, - - - - - = 0.3928
JAVA I. — Batavia, Samarang, Sarakarta: 5 duyt = 1
    stuiver, 2 stuivers = 1 dubbel, 3 dubbele = 1
    schilling, 4 schillinge = 1 Florin, - - = 0.3928
    U. S. Customs value of the Florin = $0.40.
MALACCA. — Malacca: 4 duyt = 1 stuiver, 6 stuivers =
    1 schilling, 8 schillinge = 1 Ryksdaalder, - = 0.795
MOLUCCA ISLANDS or SPICE I. — Same as BATAVIA, Java I.
PHILIPPINE I. — LUZON I. — MANILLA, &c.: 34 maravedi = 1 reál, 8 reáls = 1 Peso, - - = 1.00
SIAM. — Bangkok, &c.: 4 tical or bats = 1 tael, 100 t.
    = 1 Pecul, - - - - - = 0.617
SINGAPORE I. — Same as MALACCA.
SOOLOO ISLANDS. — Same as CHINA.
SUMATRA I. — Acheen: 4 copangs = 1 mace, 4 m. = 1
    pardo, 4 p. = 1 Tael, - - - = 4.16
    Bencoolen: 8 satellers = 1 soocoo, 4 s. = 1 Reál, = 1.10
IONIAN ISLANDS. — Cephalonia, Corfu, Ithaca,
    Paxos, Zante, &c.: 12 denari = 1 soldo, 20 s. =
    1 Lira, - - - - - =
ITALY. — 100 centesimi = 1 Lira Italiani, - = 0.187
    ECCLESIASTICAL STATES. — Standard for silver coins =
    $\frac{222}{288}$ fine; for gold coins = $\frac{900}{1000}$ fine. Relative
    values, gold to silver as 15.526 to 1.

FOREIGN MONEYS OF ACCOUNT.   a 7

| *Foreign*. | *U. States.* |

Ancona: 12 denari = 1 soldo, 20 s. = 1 Scudo, - = $1.007
100 bajochi *or* 80 bologni *or* 10 paoli = 1 Scudo.
Bologna, Ferrara: 12 denari = 1 soldo, 20 s. = 1 Lira, = 0.201
5 lire *or* 10 paoli *or* 100 baiochi *or* 500 quattrini =
1 Scudo, - - - - - = 1.007
Rome: 5 quattrini = 1 baiocho, 10 b. = 1 paolo, 10 p.
= 1 Scudo = $\frac{1}{12}$ libbra (373.78 *troy grains*) of
fine silver, - - - - - = 1.007
Sequin = $1\frac{98}{99}$ fine = $\frac{1}{100}$ libbra of fine gold.
LUCCA. — 12 denari = 1 soldo, 20 s. = 1 Lira, - = 0.243
12 denari = 1 soldo, 20 s. = 1 Scudo correnta, = 1.007
1 scudo d'oro of Lucca = $1.127.
12 denari di cambia = 1 soldo di cambia, 20 s. = 1
Scudo di cambia, - - - - = 1.008
LOMBARDY AND VENICE. — Standard for silver coins = $\frac{73}{80}$
fine ; for gold coins = $\frac{28}{30}$ fine. Relative values,
gold to silver as 16 to 1.
Bergamo, Mantua, Milan, Padua, Verona, Venice, &c. :
100 centesimi = 1 Lira Italiani, - - = 0.187
100 centesimi = 1 Lira Austriache, - - = 0.162
12 denari = 1 soldo, 20 s. = 1 Lira correnta di Milan = $\frac{73}{504\frac{3}{10}}$ marcs of fine silver = 52.531 troy
grains, - - - - - = 0.1415
1 Lira picola di Bergamo, di Verona *or* di Venice, = 0.098
1 Lira di Mantua, - - - - - = 0.058
MODENA, PARMA. — 100 centesimi = 1 Lira, - = 0.187
NAPLES. — *Naples, Salerno,* &c. : 7200 accini = 360
trapesi = 12 oncie = 1 marco or libbra = 4950.53
troy grains.
Standard for gold and silver coins = $\frac{925}{1000}$ fine, each.
Relative values, gold to silver as 15.373 — to 1.
10 grani = 1 carlino, 10 c. = 1 Ducato = $\frac{4}{67}$ Neapolitan libbra of fine silver = 295.55+ troy
grains, - - - - - = 0.796
1¼ ducati = 1 scudo or crown.
SARDINIA : 4608 grani of 24 granottini, each = 1152
carati = 192 denari = 8 oncie = 1 marco = 3795
troy grains.
Standard for gold and silver coins = $\frac{87}{96}$ fine, each.
Relative values, gold to silver as 15.545+ to 1.
100 centesimi = 1 Lira Italiani, - - = 0.187
12 denari = 1 soldo, 20 s. = 1 Lira di Sardinia =
$\frac{1}{29}$ marco of fine silver = 130.86 troy grains, - = 0.3525
Genoa: 5¾ lire fuori banco = 1 Pezza, - - = 0.8854

FOREIGN MONEYS OF ACCOUNT.

Foreign.	U. States.
4¾ lire di cambio = 1 scudo *of Exchange*, - -=	$0.7187
10 $\frac{81}{100}$ lire moneta buona = 1 Scudo d'oro, - =	1.6663

TUSCANY. — *Florence* — moneta buona: 12 denari = 1 soldo, 20 soldi = 1 Lira, - - - = 0.1566
12 denari di ducato = 1 soldo di ducato, 20 s. = 1 Ducato *or* scudo correnta,   -   - -= 1.0962
7 lire = 1 ducato. 7¼ lire = 1 scudo d'oro.
Leghorn. — moneta lunga: 12 denari di lira = 1 soldo di lira, 20 soldi di lira = 1 Lira, - = 0.15
12 denari di pezza = 1 soldo di pezza, 20 soldi di pezza = 1 Pezza, -   -   -   -   -= 0.9004
6 lira = 1 pezza.
7 lire moneta buona = 1 Scudo corrente, - = 1.0962
7¼ lire moneta buona = 1 Scudo d'oro, - -= 1.1745

JAPAN. — *Matsmay, Miaco, Nangasaki, Osaca, Yedo,* &c.:
10 cash = 1 candarine, 10 c. = 1 mace, 10 m. = 1 Tael, - - - - - = 1.4074

MADEIRA ISLANDS. — 1000 reas = 1 Milrea, -= 1.00
MALTA I. — 20 grani = 1 taro, 12 t. = Scudo, - = 0.406
6 piccioli = 1 carlino, 2 c. = 1 taro, 12 t. = 1 Scudo, - - - - - -= 0.406
2½ scudi = 1 Pezza, - - - - = 1.015

MAURITIUS I. — *Port Louis,* &c.: 100 cents = 1 Dollar, - - - - - -= 0.9353
20 sols = 1 Livre colonial = ⅒ Franc of France,   = 0.09353
10 livres = 1 Dollar.

MEXICO. — *Acapulco, Tampico, Vera Cruz,* &c.:
6 grani = 1 cuarto, 2 c. = 1 medio, 2 m. = 1 Reál, = 0.125
8 reáls = 1 Peso, - - - - = 1.00

MOROCCO. — *Fez, Mogadore,* &c.: 24 fluce = 1 blankeel, 4 b. = 1 once, 10 onces = 1 Mitkul, = 0.7407

NORWAY. — *Bergen:* 16 skillinge = 1 mark, 6 m. = 1 Ryksdaler, - - - - -= 0.5252
2 Ryksdalers = 1 Speciesdaler.
Christiania, &c.: 12 pfennige = 1 skilling, 6 s. = 1 ort, 4 orte = 1 Ryksdaler = 1 mark banco of Hamburg, - - - - -= 0.3501
3 Ryksdaler = 1 Speciesdaler.

PERSIA. — *Bushire,* &c.: 5 denari = 1 kasbeque, 10 k. = 1 shafree, 2 s. = 1 mamoode, 2 m. = 1 abasso, 50 a. = 1 Toman, - - - - = 2.233
2 kasbequi = 1 denaro-biste.

PORTUGAL: Standard for silver coins = $\frac{21}{88}$ fine — for gold coins = $\frac{1}{12}$ fine. Relative values, gold to silver as 15.356+ to 1.

FOREIGN MONEYS OF ACCOUNT.   a 9

        Foreign.   *U. States*
Method of writing and reading quantities: *Ex.* —
  rs. 5 : 600 ⊕ 750 = 5,600 milreas and 750 reas.
1000 reas = 1 Milrea = the silver coroa = $\frac{122}{135}$
  marco of fine silver = 415.435 troy grains, - = $1.119
1000 reas = 1 Milrea current — fluctuating, about
  = $0.96.
1 Milrea, paper — fluctuating, about = $0.81.
480 reas = 1 Crusado.
PRUSSIA. — Standards and relative values, same as
  given under GERMANY.
  Berlin, Brandenburg, Dantzic, Potsdam, Magdeburg,
    Stetin, &c. :
    12 pfennig = 1 gute groschen, 24 g. = 1 Rixdollar
      or Thaler current = $1\frac{1}{20}$ specie thaler,   -   = 0.7287
    1¼ Rixdollar = 1 Thaler banco = 0.91\frac{7}{80}$.
    1¾ florins = 1 Thaler specie,   -   -   - = 0.694
  Cologne: 12 hellers = 1 albus, 80 a. = 1 Rixdollar
    = $\frac{3}{20}$ Convention pistole = 1½ florins d'or, - = 0.6033
    78 albus = 1 Rixdollar current.
    120 fettmangen = 90 kreuzer = 30 groschen = 20
      blafferts = 3½ Cologne florins = 2 heron florins
      = 1¼ rader florins.
  Aix la Chapelle, Crevelt, Elberfeldt, &c. :
    4 pfennig = 1 kreuzer, 60 k. = 1 florin, 1¾ f. = 1
      Thaler,   -   -   .   -   -   = 0.694
  Brunswick: 8 pfennig = 1 marien-groschen, 36 m.
    = 1 Thaler,   -   -   -   -   -   = 0.694
  Konigsberg: 6 pfennig = 1 schilling, 3 s. = 1 gros-
    chen, 30 groschen = 1 florin, 3 f. = 1 Thaler,   = 0.694
    8 specie gute groschen of Berlin = 30 groschen of
    Konigsberg.
PRINCE OF WALES I. — 10 pice = 1 copang, 10 c.
  = 1 Dollar,   -   -   -   -   -   = 1.00
PROVINCES OF NEW BRUNSWICK, NOVA SCOTIA, NEW-
  FOUNDLAND, AND THE CANADAS :
  4 farthings = 1 penny, 12 p. = 1 shilling, 20 s. =
    1 Pound,   -   -   -   -   -   = 4.00.
RUSSIA. — Standard for silver coins = ⅞ fine, for gold
  coins = $\frac{11}{12}$ fine. Relative values, gold to silver
  as $15\frac{9}{34}$ to 1.
  Archangel, Cronstadt, Helsingfors, Odessa, Revel, Se-
    vastopol, St. Petersburg, &c. :
    10 kopecs = 1 grieven, 10 g. = 1 Ruhlyu (*ruble*)
      = $\frac{3}{68}$ funt of fine silver = 278.47 troy grains, = 0.75

FOREIGN MONEYS OF ACCOUNT.

Foreign.	U. States.
2 denushkas *or* 4 polushkas = 1 kopec. 33⅓ altins = 1 ruble.	
Riga. — Same as St. Petersburg, also —	
30 groschen = 1 florin, 3 f. = 1 Rixdollar. 1 Albertus dollar, - - - - - =	$1.00
SARDINIA I. — 100 centesimi = 1 Lira Italiani, =	0.187
1 Lira di Sardinia, - - - - =	0.35¼
SICILY I. — Standards of purity and relative values, same as NAPLES.	
6 picioli = 1 grano, 20 g. = 1 taro, 30 t. = 1 Oncia = $\frac{13}{67}$ Neapolitan libbra of fine silver, - - =	2.388
8 picioli = 1 ponti, 15 p. = 1 taro, 10 t. = 1 ducato.	
6 tari = 1 fiorino or florin, 2 f. = 1 scudo, 2½ s. = 1 oncia.	
SPAIN. — Standard for silver coins since 1786, peso and ½ peso = $\frac{13}{18}$ fine; peseta, reál and ⅛ reál = $\frac{33}{38}$ fine; for gold coins = ⅞ fine, except the coronilla (*gold dollar*) = $\frac{153}{182}$ fine. Relative values, since 1786, gold to silver as 16.39 to 1.	
Reál vellón = $\frac{1}{20}$ peso duro, - - - =	0.05
Reál de plata nuevo = $\frac{1}{10}$ peso duro, - - =	0.10
Reál de plata Mexicana = ⅛ peso duro, - - =	0.125
Reál de plata antiquas = $\frac{49}{125}$ peso duro, - - =	0.0943
Reál d'Alicant = $\frac{32}{125}$ peso duro, - - =	0.0754
Reál de Valencia = $\frac{24}{125}$ peso duro, - - =	0.0566
Reál currante de Gibraltar = $\frac{1}{12}$ peso duro, - =	0.0835
Reál de Catalonia = $\frac{48}{595}$ peso duro, - - =	0.0808
Reál ardita de Catalonia = $\frac{32}{595}$ peso duro, - =	0.0539
8 reále de plata antiquas = 1 Piastre or peso of exchange = $\frac{64}{85}$ peso duro, - - - =	0.7543
Peso duro = $\frac{163}{153}$ marco of fine silver = 371.9 troy grains, - - - - - =	1.0018
40 dineri = 16 comadi = 8 blanci = 4 maravedi = 2 ochavi = 1 quarto, 4¼ quarti = 1 sualdo, 2 s. = 1 Reál.	
Alicant: 34 maravedi = 1 reál, 10 r. = 1 Piastro, =	0.7543
Barcelona, Tortosa: 2 malli = 1 dinero, 12 d. = 1 sualdo, 20 s. = 1 Libra = 10 reále ardita de Catalan, - - - - =	0.5388
Bilboa, Carthagena, Madrid, Malaga, Santander, Toledo:	
34 maravedi = 1 reál, 15 r. = 1 Peso sencillo.	

Foreign.	U. States.

$15\frac{1}{17}$ reále = 1 peso de plata or Piastre, = $0.7543
4 piastres = 1 doubloon de plata or pistole of exchange.
Cadiz, Sevilla: 34 maravedi *or* 16 quarti = 1 Reál, = 0.0943
8 reále = 1 Piastre, 10½ reále = 1 Peso duro.
Gibraltar: 34 maravedi = 1 reál, 9 r. = 1 Piastre, = 0.7543
12 reále = 1 Peso duro.
Valencia: 12 dineri = 1 sualdo, 2 s. = 1 reál, 10 r.
  = 1 Libra, = 0.5657
1½ libra = 1 Piastre or peso of account, = 0.7543
2¼ dineri = 1 reál, 10 r. = 1 Peso duro, = 1.0018
SWEDEN. — Standard for gold coins (ducats, multiples and fractions) = $\frac{23}{24}$ fine; for silver coins = $\frac{19}{24}$ fine. Relative values, gold to silver as 14.692 to 1.
Carlscrona, Gefle, Gottenburg, Stockholm, &c.:
  12 rundstycken *or* öre = 1 skilling, 48 s. = 1 Riksdaler = $\frac{3}{25}$ mark of fine silver = 393.68 troy grains, = 1.0604
100 centimes *or* skillings = 1 Riksdaler, = 1.0604
SWITZERLAND. — 1 Livre de Suisse, of the convention of 1814, = 1½ *livres tournois* of France = $\frac{19}{24}$ francs of France, = 0.2771
Berne, Basle, Lausanne, Lucerne, Pay de Vaud:
  10 rappen = 1 batz, 10 batzen = 1 Livre de Suisse.
  12 deniers = 1 sou, 20 sols de Suisse = 1 Livre.
  10 rappen = 1 batz, 15 b. = 1 Florin *or* Guilder, = 0.4156
  8 hellers = 1 kreuzer, 60 k. = 1 Florin.
Geneva: 12 deniers = 1 sou, 20 s. = 1 Livre = 3½ florins petite monnie = 1⅔ *francs* of France, = 0.3117
3 livres = 1 Ecu *or* Patagon, = 0.8313
Neufchatel: 100 rappen = 1 Franc *or* Livre de Suisse.
  12 deniers = 1 sol, 20 s. = 1 Livre tournois de Neufchatel = 2¼ livers foible, = 0.2628
St. Gaul: 480 heller = 240 pfennig = 60 kreuzer = 15 batzen = 10 skilling = 1 Florin *or* Guilder.
  1 florin current = 2⅓ *francs* of France, = 0.4365
  1 florin specie, = 0.5187
Zurich: 60 kreuzer of 8 hellers each, *or* 16 batzen of 10 augsters each, *or* 40 skillings of 12 hellers each
  = 1 Guilder or Florin, = 0.4365
TRIPOLI. — 100 paras = 1 Piastre *or* Ghersch, ghersch of 1832, = 0.10
TUNIS. — *Tunis, Biserta, Susa,* &c.: 2 burbine = 1

FOREIGN MONEYS OF ACCOUNT.

Foreign.   *U. States.*

asper, 52 a. or 16 carobas = 1 Piastre,* piastre of 1838, - - - - - - = $0.128

TURKEY. — *Constantinople :* 3 aspers = 1 para, 40 p. = 1 Piastre *or* Ghersch.*

With the Dutch, French and Venetians, 100 aspers = 1 Piastre.

With the English and Swedes, 80 aspers = 1 Piastre. 500 piastres = 1 chise; 30,000 piastres = 1 kitz; 100,000 piastres = 1 juck.

Smyrna : 40 paras *or* medini = 1 Piastre *or* Gooroosh. 12 tomans = 1 Piastre or Gooroosh.

West Indies.

CUBA I. — *Cardenas, Cienfuegos, Havana, Matanzas, Mariel, Nuevitas, Porto Principe, Sagua la Grande, St. Jago,* &c. :
   34 maravedi = 1 reál, 8 r. = 1 Peso, - - = 1.00

HAYTI I. — *Aux Cayes, Cape Haytien, Port au Prince, San Domingo,* &c. :
   100 centesimi = 1 Dollar *or* Peso duro = 11 esculini, - - - - - = 1.00
   1 dollar Haytien currency = 7½ esculini, - - = 0.66

PORTO RICO I. — *Guayama, Mayaguez, St. Johns, Ponce,* &c. :
   34 maravedi = 1 reál, 8 r. = 1 Peso, - - = 1.00

BRITISH ISLANDS. — *Anguilla, Antigua, Barbuda, Dominica, Grenada, Montserrat, Nevis, St. Kitts, St. Lucia, St. Vincent, Tobago, Tortola, Trinidad, Virgin Gorda :*
   4 farthings = 1 penny, 12 p. = 1 shilling, 20 s. = 1 pound, - - - - - - = 2.222

Nassau and the Bahamas generally :
   4 farthings = 1 penny, 12 p. = 1 shilling, 20 s. = 1 Pound, - - - - - = 2.485
   Pound of Turks I. - - - - - = 3.00

BARBADOES I. — *Bridgetown,* &c. : 1 Pound, - = 3.20

JAMAICA I. — *Falmouth, Kingston, Morant Bay, Savannah la Mar,* &c. :
   4 farthings = 1 penny, 12 p. = 1 shilling, 20 s. = 1 Pound, - - - - - - = 3.00

---

* The coins of the Turkish government, owing to frequent and oft-repeated deterioration by enactments, have no definable standard value whatever. Bills of exchange on Turkey are usually drawn in Spanish dollars. The value of the silver piastre of Turkey, of full weight, of 1775, is $0.446; of that minted in Tunis in 1787, $0.259; of that of Turkey of 1818, $0.182, and of that of 1836, $0.128, while that issued only a few years since, is worth, intrinsically, but about 4 cents.

FOREIGN MONEYS OF ACCOUNT. *a* 13

Foreign.   *U. States.*

DANISH ISLANDS.—*Santa Cruz, St. John, St. Thomas, St. Bartholomew:*
   12 skillings = 1 bit, 8 b. = 1 Ryksdaler,   — $0.64
   100 cents = 1 Ryksdaler.
   12½ bits = 1 Spanish dollar.

DUTCH ISLANDS.—*Saba, St. Eustatius, St. Martin:*
   6 stuivers = 1 reál, 8 r. = 1 Piastre,   -   -— 0.73
   11 reáls or Esculins = 1 Spanish dollar.

FRENCH ISLANDS.—*Deseada, Guadeloupe, Mariegalante, Martinique:*
   12 deniers = 1 sol, 20 s. = 1 Livre = ⅜ livre tournois, - - - - - - — 0.1232
   4 farthings = 1 penny, 12 p. = 1 shilling, 20 s. = 1 Pound, - - - - -— 2.222

LITTLE ANTILLES, generally, Same as Mexico.

FOREIGN LINEAR AND SURFACE MEASURES REDUCED TO UNITED STATES.

Foreign.	*U. States.*
ABYSSINIA. — *Massuah :* 8 robi = 1 derah *or* pic, =	0.682 yard.
ALGIERS. — 10 decimetres = 1 metre, - -=	1.094 "
8 robi = 1 pic. Pic, *Moorish,* for linens, - =	0.519 "
Pic, *Turkish,* for silks, &c., -=	0.692 "
ARABIA. — 1 kassaba = 12.31 ft. Mile, - =	1.22 miles.
Aden : 8 robi = 1 yard *or* pic, - - -=	0.95 yard.
Jidda : 8 robi = 1 pic, - - - =	0.743 "
Mocha : 8 gheria = 1 covid. Covid (*land*), -=	1.58 feet.
Covid (*for iron, &c.*), - - =	2.25 "
8 robi = 1 gez, - - - =	0.604 yard.
AUSTRIA. — (*Imperial, or legal and general*) :	
Vienna, Trieste, Prague, Lintz, &c. :	
12 zoll = 1 fus, - - - - =	1.037 feet.
29½ zoll = 1 elle, - - - -=	0.852 yard.
6 fus = 1 klafter, 4000 k. = 1 meile, - =	4.712 miles.
10 fus = 1 ruth (*builders'*), - - -=	10.37 feet.
3 metzen = 1 joch, - - - =	1.422 acres.
(*Special and local*) —	
UPPER AUSTRIA. — *Lintz, &c. :* 1 elle, - -=	0.874 yard.
BOHEMIA. — *Prague, &c. :* 2 fus — 1 elle, - =	0.65 "
4 elle = 1 dumplachter, - - -=	2.598 "
HUNGARY. — 1 fus = 1.037 feet. 1 elle, - =	0.874 "
MORAVIA. — 2⅔ fus = 1 elle, - - -=	0.865 "
AZORE ISLANDS. — Same as LISBON (*Portugal*).	
BALEARIC ISLANDS. — 3 pic *or* 4 palma = 1	
vara, 2 vara = 1 cana.	
MAJORCA. — 1 cana, - - =	1.711 yards.
MINORCA. — 1 cana, - - -=	1.754 "
BELGIUM. — 10 streep = 1 duim, 10 d. = 1 palm,	
10 p. = 1 el. = 1 *metre* of France, - =	1.093 yards.
10 el = 1 roed, 100 r. = 1 mijl, - -=	0.621 mile.
2½ fus = 1 auno.	
Antwerp : Aune *for cloths,* - - - =	0.749 yard.
Aune *for silks,* - - -=	0.761 "
Brussels : 1 auno = 0.761 yards. Vaem, - =	2.— "

FOREIGN LINEAR AND SURFACE MEASURES. *a* 15

Foreign.	*U. States.*
Mechlin : 1 aune, — =	0.753 yard.
BERMUDAS I. — Same as GREAT BRITAIN.	
BOURBON I. — 3 pied = 1 aune, — =	1.298 "
BRAZIL. — 12 pollegada = 1 pe, 5 pes = 1 passo, 52 passi = 1 estadio, 24 estadi = 1 milha, 3 milhe = 1 legoa, — =	3.836 miles.
8 pollegada = 1 palmo, 5 palmi = 1 vara, 2 vare, *or* 3½ covadi, *or* 1⅓ passi = 1 braça, =	7.214 feet.
1 geira, — =	1.428 acres.
Bahia, Rio Janeiro : 3 palmi = 1 covado, — =	0.713 yard.

Central and South America.

BALIZE, BOLIVIA, BUENOS AYRES, CHILI, EQUADOR, GUATIMALA, NEW GRANADA, PERU, URUGUAY, VENEZUELA, YUCATAN :
Nomenclatures and legal values, same as CASTILE (*Spain*).
GUIANA. — *Berbice, Demerara, Essequibo, Surinam:* Same as HOLLAND.
Cayenne. — Same as FRANCE.

CANARY I. — 12 onza = 1 pie, 3 p. = 1 vara, =	0.920 yard.
2 vara = 1 braza, — =	5.522 feet.
52 braza cuadrada = 1 celemin, 12 c. = 1 fanegada, — =	0.5 acre.
CANDIA I. — 8 robi = 1 pic, — =	0.697 yard.
CAPE COLONY. — Same as GREAT BRITAIN.	
CAPE VERDE I. — Same as LISBON (*Portugal*).	
CHINA. — 10 fan = 1 tsun *or* punt, 10 tsun = 1 kong-pu *or* chik, 10 kong-pu = 1 cheung, 10 cheung = 1 yan, 18 yan = 1 li, — =	0.346 mile.
Chik (*mathematical*) = 1.094 ft. Chik (*engineers'*), =	1.058 feet.
Chik (*tradesmen's*) = 1.218 ft. Kong-pu, — =	1.014 "
1½ chik (*engineers'*) = 1 thuoc, 3⅙ thuoc = 1 po, — =	5.025 "
10 punts = 1 covid *or* cobre, 1¾ c. = 1 thuoc (*mercers'*), — =	0.711 yard.
Pekin : 10 chik (*math.*) = 1 cheung, — =	10.937 feet.
CYPRUS I. — 8 robi = 1 pic, — =	0.696 yard.
DENMARK. — 24 tomme *or* 2 fod = 1 aln, — =	0.688 "
3 aln = 1 favn, 1¾ f. = 1 rode, 2400 r. = 1 miil, =	4.681 miles.
96 album *or* 8 skiepper = 1 toende, — =	5.45 acres.
EGYPT. — 2 derah = 1 fedan, 3 f. = 1 gasab, =	12.67 feet.
8 rob = 1 pic, — =	0.74 yard.

FOREIGN LINEAR AND SURFACE MEASURES.

Foreign.	U. States.
1 fedan al risach, =	4.—— acres
Alexandria, Rosetta: 1 pic stambuli, =	0.733 yard.
Pic for muslins, &c., =	0.686 "
Pic for cloths, =	0.613 "

FRANCE. — 100 centimètres or 10 decimètres = 1
métre, = 1.094 "
100 mètres or 10 decamètres = 1 hectomètre, = 19.883 rods.
100 hectomètres or 10 kilomètres = 1 myria-
mètre, = 6.214 miles.
100 square mètres = 1 are, 100 a. = 1 hectare, = 2.471 acres.
3 6/10 pied metrique = 1 aune = 47¼ inches, = 1.312 yards.

GERMANY. — BADEN (legal): 20 zoll or 2 fus =
1 elle, = 0.656 "
5 elle = 1 ruthe = 3 mètres of France, = 3.281 "
2 stunden = 1 meile, = 5.524 miles.
1 jauchart = 0.82 acre. 1 morgen, = 0.889 acre.
Manheim: 1 fus = 0.952 ft. 1 elle, = 0.610 yard.
BAVARIA (legal): 120 zoll or 10 fus = 1 ruthe, = 9.575 feet.
2400 ruthe = 1 meile, = 4.352 miles.
34¼ zoll = 1 elle, = 0.911 yard.
1 jauchart or morgen, = 0.841 acre.
5 cubic fus = 1 klafter = 110.62 cubic feet.
Augsburg: 2 fus = 1 elle, = 0.648 yard.
1 elle (mercers'), = 0.666 "
Nuremberg: 2¼ fus = 1 elle, = 0.718 "
HANOVER (legal): 12 zoll = 1 fus, = 0.943 foot.
2 fus = 1 elle = 0.638 yard. 8 e. = 1 ruthe, = 15.328 feet.
1462½ ruthe = 1 meile, = 4.246 miles.
2 vierling = 1 morgen, = 0.647 acre.
Bremen: 24 zoll or 2 fus = 1 elle, = 0.633 yard.
6 fus = 1 klafter, 2⅔ k. = 1 ruthe, = 15.188 feet.
20000 Rhineland fus = 1 meile, = 3.896 miles.
120 square ruthe = 1 morgen, = 0.636 acre.
1 reif = 96.52 cub. ft. 1 faden = 61.6 cub. ft.
Emden, Osnaburg: 2¼ fus = 1 elle, = 0.698 yard.
HESSE CASSEL. — 24 zoll or 2 fus = 1 elle, = 0.623 "
14 fus = 1 ruthe, = 13.088 feet.
1 klafter = 126.089 cubic feet.
HESSE DARMSTADT (legal): 100 zoll or 10 fus =
1 klafter = 2½ mètres of France, = 8.202 feet.
32 zoll = 1 elle, = 0.875 yard.
400 square klafter or 4 viertel = 1 morgen, = 0.618 acre.
Frankfort: 12 zoll = 1 fus or werkschuh, = 0.934 foot.
2 fus = 1 elle, 2 e. = 1 stab, = 1.245 yards.
10 feldfus = 1 ruthe, = 11.672 feet.

Foreign.   *U. States.*

HOLSTEIN. — *Hamburg, Altona:*
   24 zoll *or* 6 palm *or* 2 fus = 1 elle, - - = 0.6266 yard.
   3 elle = 1 klafter, 2¼ k. = 1 marschruthe, = 13.159 feet.
   2⅔ klafter (16 fus) = 1 geestruthe, - = 5.013 yards.
   24000 Rhineland fus (2000 R. ruthe) = 1 meile, - - - - = 4.68 miles.
   1 Brabant elle *for woollens*, - - = 0.761 yard.
   600 square marschruthe = 1 morgen, - = 2.385 acres.

Lubec: Denominations and relative values, same as at *Hamburg.* — 1 elle, - - := 0.63 yard.

MECKLENBURG. — *Rostock, &c.* — Same as *Hamburg.*

SAXONY. — *Dresden, Leipsic:* 12 linie = 1 zoll, 12 zoll = 1 fus, 2 fus = 1 elle, - - = 0.618 yard.
   2 elle = 1 stab, 4 stab = 1 ruthe, - -= 4.943 "
   3 elle = 1 klafter, - - - = 5.561 feet.
   1500 ruthe = 1 meile, - - -= 4.213 miles.
   300 square ruthe = 1 acker, - - = 1.515 acres.

Freyburg: 2 fus = 1 elle, 5 e. = 1 ruthe, -= 9.619 feet.

OLDENBURG. — 24 zoll *or* 2 fus = 1 elle, - = 0.648 yard.
   9 elle = 1 ruthe, 1850 r. = 1 meile, - -= 6.133 miles.

GREAT BRITAIN. — Same as United States.

GREECE. — *Patras:* 8 robi = 1 pic *for silks*, -= 0.694 yard.
   1 pic *for woollens, &c.*, = 0.75 "

HOLLAND (*legal*): 10 streep = 1 duim, 10 d. =
   1 palm, 10 p. = 1 el = 1 *metre* of France, = 1.093 "
   10 el = 1 roed, 100 r. = 1 mijl, - -= 0.621 mile.

Previous to 1820 — 2⅓ fus = 1 el, - - = 0.747 yard.
   El *of Flanders,* - - - -= 0.776 "
   HAGUE — *Brabant* el, - - - = 0.761 "

India and Malaysia or East Indies.

AN-NAM. — Same as CHINA.

BIRMAH. — 4 taim = 1 sadang, 7 s. = 1 bambou, = 4.208 yards.

CEYLON I. — *Colombo:* 5 palmi = 1 covid, - = 0.516 "

HINDOSTAN. — *Bombay:* 2 tussoo = 1 gheria, 8 g.,
   = 1 haut *or* covid, - - - -= 0.503 "
   1½ haut = 1 guz, - - - = 0.755 "

Calcutta: 3 jaob = 1 angulla, 3 a. = 1 gheria,
   8 g. = 1 haut *or* covid, 2 h. = 1 ghes *or* guz, = 1.— "
   3 palgat = 1 hand, 5 h. = 1 cubit, - -= 1.25 foot.
   3 cubits = 1 corah, 1728 c. = 1 coss, - = 1.227 miles.

Goa: 12 pollegada = 1 pe, - - -= 1.082 feet.

FOREIGN LINEAR AND SURFACE MEASURES.

Foreign.	U. States.
24¾ pollegada = 1 covado avantejado, =	0.744 yard.
13¼ pollegada = 1 terca, 3 t. = 1 vara, -=	1.203 "
Madras: 8 gheria = 1 covid, =	0.515 "
1 cassency *or* cawney, -=	1.32 acres.
Massulipatam, 2 palm = 1 span, 3 s. = 1 cubit, =	1.594 feet.
Mysore, Seringapatam: 8 gerah = 1 haut, 2 h. = 1 gugah, =	1.072 yards.
Pondicherry: 8 gheria = 1 haut *or* covid, -=	0.5 "
Surat: 84 tussoo *or* 20 wiswüsa = 1 wüsa, =	2.712 "
18 tussoo = 1 cubit *or* haut *for matting*, - =	0.581 "
Tatta: 16 garca = 1 guz, -=	0.943 "
Tranquebar, Scrampore (*legal*): same as DENMARK,	
JAVA I. — *Batavia*: 8 gheria = 1 covid, cubit *or* cl, =	0.75 "
1 fus (*Rhenish*), =	1.03 feet.
MALACCA. — *Malacca*: 8 gheria = 1 covid, - -=	0.5 yard.
8 covid = 1 jumba, =	12.— feet.
PHILIPPINE I. — *Luzon I.* — *Manilla.* — Same as CADIZ, *Spain*.	
SIAM. — 12 nion = 1 keub, 2 k. = 1 sok, - -=	0.525 yard.
2 sok = 1 ken, 2 k. = 1 vouah, 20 v. = 1 sen, 40 s. = 1 jod, 25 jod = 1 ročneng, =	2.388 miles.
Bangkok: 8 gheria = 1 covid, -=	1.5 feet.
SINGAPORE I. — 8 gheria = 1 hasta, =	1.5 "
SUMATRA I. — 4 tempoh *or* 2 jankal = 1 etto, 2 etto = 1 hailoh, -=	1.03 yard.
IONIAN ISLANDS. — *Cephalonia, Corfu, Ithaca, Paxos, St. Maura, Zante*, &c. :	
12 onué = 1 pic, 5 pes = 1 passo, - =	5.455 feet.
1 braccio *for silks*, -=	0.705 yard.
1 braccio *for woollens*, =	0.755 "
1 moggio (*linear*), -=	2.4 miles.
36 inches *or* 3 feet = 1 yard, =	1.— yard.
ITALY. — LOMBARDY AND VENICE:	
Government and Customs Measure —	
10 atome = 1 dito, 10 d. = 1 palmo, 10 p. = 1 metro *or* braccio = 1 *metre* of France, -=	3.281 feet.
1000 metre = 1 miglio, =	0.621 mile.
100 square metre = 1 tavola, 100 t. = 1 tornatura, -=	2.471 acres.
Special and local —	
Venice: 2 palmi = 1 braccio, 2½ b. = 1 passo, =	5.699 feet.
1⅕ passi = 1 pertica.	
4½ pede — 1 chebbo, 1¼ c. = 1 cavezzo, -=	6.845 "
1 passo, *geometrical*, =	4.559 "
1 braccio *for woollens*, -=	0.739 "

FOREIGN LINEAR AND SURFACE MEASURES.   *a* 19

```
    Foreign.                                        U. States.
  1 braccio, for silks,             -    -    -  =  0.693  feet.
NAPLES. — 5 minuto = 1 oncia, 12 o. = 1 palmo, 8
   palmi = 1 canna,      -     -     -     -   =   6.92   "
  7½ palmi = 1 passo or pertica, 8 pertica = 1
   catena, 116⅔ catene = 1 miglio,    -      =   1.147 miles.
  900 square passi = 1 moggio,  -    -    - =    0.87  acre.
  1 braccio (2⅔ palmi in theory),    -    -  =   0.764 yard.
SARDINIA. — Genoa: 8 oncie = 1 pie, 10 p. or 12
   palmi = 1 canna (surveyors'),      -     - =   9.715 feet.
  2¼ palmi = 1 braccio,     -     -     -    =   0.63  yard.
  9 palmi = 1 canna picolo,    -     -     - =   2.429  "
  10 palmi = 1 canna, for linens.
  12 oncia = 1 pie liprando.
Nice: 12 oncia = 1 palmo. 25 oncia = 1 raso,  =  0.600  "
Turin: 8 oncia = 1 pie manual,    -     -    =   1.19  feet.
  12 oncia = 1 pie liprando.
  14 oncia = 1 raso,    -     -     -     - =    0.649 yard.
  5 pie manual = 1 tesa.
  6 pie lip. = 1 trabucco, 2 t. = 1 pertica.
STATES OF THE CHURCH. — Ancona: 1 braccio,   =   0.704  "
  10 pie = 1 pertica,      -     -     -    =  13.438 feet.
Rome: 10 decime or 5 minuto = 1 oncia, 16 o.
   = 1 pie, 5 piede = 1 passo,    -     -  =   4.884 feet.
  5 linea = 1 parto, 24 p. = 1 palmo, 8 palmi =
   1 canna,     -     -     -     -     -  =   2.176 yards.
TUSCANY. — Leghorn, Florence, Pisa:
  12 denari or 3 quattrini = 1 soldo, 20 soldi or 2
   palmi = 1 braccio, 4 b. = 1 canna,  -    =   2.552  "
  3 braccia = 1 passo, 2 p. = 1 cavezzo,  - =  11.484 feet.
  5 braccia = 1 pertica, 566⅔ p. = 1 miglio, =  1.027 miles.
JAPAN. — 5 Kupera sasi = 1 ink,    -     -  =  2.072 yds.
  1¼ sasi = 1 k. sasi, 2¼ sasi = 1 ikje, -  =   2.32  feet.
MALTA I. — 12 oncie = 1 palmo, 8 palmi or 7½
   piede = 1 canna,     -     -     -     - =   2.275  "
MADEIRA I. — Standard same as LISBON, Port-
   ugal.
MAURITIUS I. — Standard same as GREAT BRIT-
   AIN.
MEXICO. — Same as CADIZ, Spain.
MOROCCO. — Mogadore: 1 cadec,    -     -   =   1.695 feet.
  1 covado = 1.654 feet. 1 pic,    -     -  =   0.723 yard.
NORWAY. — Same as DENMARK.
PERSIA. — 1 archin arisch,   -     -     - =   1.063  "
  1 archin schah,      -     -     -     - =   0.874  "
  1 gueza (royal) = 3 1/10 feet. 1 gueza (com-
   mon),    -     -     -     -     -     - =   0.692  "
```

Foreign.	U. States.
1 monkelzer,	= 2.351 feet.
Bushire: 1 guz,	= 0.557 yard.

PORTUGAL. — *Lisbon, St. Ubes, &c.:*
80 pollegada, *or* 10 palmi de craveira, *or* 6⅔ pes, *or* 3½ covado, *or* 2 vara, *or* 1⅓ passo = 1 braça, = 7.214 feet.
780 pes = 1 estadio, 8 estadi = 1 milha, = 1.279 miles.
3 milha = 1 legua.
8 pollegada = 1 palma, 3 p. = 1 covado, = 0.722 yard.
24¾ pollegada = 1 covado avantejado, = 0.744 "
13⅓ pollegada = 1 terca, 3 t. = 1 vara, = 1.203 "
Oporto: 3 palma = 1 covado, = 0.707 "

PRUSSIA (*legal since* 1820):
12 zolle = 1 fus (Rhein-fus), = 1.03 feet.
10 zolle = 1 land-fus, 10 land-fus *or* 12 Rhine-fus = 1 ruth, 2000 r. = 1 meile, = 4.68 miles.
25½ zolle (Rhein-zolle) = 1 elle, = 0.7293 yard.
180 square ruthe = 1 morgen.
Dantzic (*special*): 75 arn = 1 seil, = 47.062 "
Konigsberg: 1 elle, = 0.628 "

RUSSIA (*legal for the Empire*):
16 verschok = 1 archine, = 0.777 "
3 archines *or* 7 feet = 1 sachine, = 7.— feet.
500 sachines = 1 verst, = 0.663 mile.
2400 square sachine = 1 deciatine, = 2.7 acres.
CRIMEA. — *Sevastopol, &c.:* 1 halebi, = 0.799 yard.

SARDINIA I. — 12 oncia = 1 palma, = 0.286 "
22 oncia = 1 pie, = 1.571 feet.
25½ oncia = 1 raso, = 0.6 yard.
12 palmi = 1 trabucco. 10 palmi = 1 canna, = 2.87 "

SICILY I. — *Messina:* 8 palmi = 1 canna, = 2.311 "
Palermo: 8 palmi = 1 canna, = 2.07 "

SPAIN. — *Alicant:* 9 pulgada = 1 palmo, 4 palmi = 1 vara, 2 vara = 1 braza, = 1.677 "
12 pulgada *or* 1½ palmi = 1 pie, = 0.833 foot.
Barcelona: 4 palmi = 1 vara *or* matja-cana, = 0.849 yard.
2 vara = 1 cana, = 5.094 feet
Cadiz (*Standard of* CASTILE): — 6 pulgada = 1 sesma, 2 s. = 1 pie *or* tercia, 3 pie = 1 vara, = 2.782 "
5 pie = 1 passo.
2 octava = 1 quarta *or* palma, 4 q. = 1 vara, = 0.928 yard
12 pulgada = 1 pie, 1½ pie = 1 codo, 2 c. = 1 vara, 2 v. = 1 estado, braza, brazada *or* tocsa, 2 e. = 1 cuerda, 2 c. = 1 cordel, 500 cordele = 1 legua, = 4.215 miles.

Foreign.		*U. States.*
192 vara cuadrada = 1 quartillo, 4 q. = 1 celemin, $7\frac{1}{5}$ c. = 1 arançada, $1\frac{3}{4}$ a. = 1 fanegada, - - - - -	=	1.587 acres.
50 fanegada = 1 yugada.		
Corunna, Ferrol: 4 palma *or* 8 octava = 1 vara,	=	0.928 yard.
Gibraltar. — As at *Cadiz*; also, as in *Great Britain.*		
Malaga. — Same as *Cadiz.*		
Santander: 8 octava *or* 4 palma = 1 vara, -	=	0.913 "
Valencia: 9 onze = 1 palmo, $1\frac{1}{3}$ p. = 1 pie,	=	0.992 feet.
3 pie = 1 vara, $2\frac{1}{4}$ v. = 1 braza reale, -	=	2.232 yards.
SWEDEN. — 12 tum = 1 fot, 2 f. = 1 aln, -	=	0.648 "
3 aln = 1 famn, $2\frac{2}{3}$ famn = 1 stang, -	=	15.553 feet.
2250 stang = 1 mil, - - - -	=	6.627 miles.
4 kappland = 1 fjerding, 4 f. = 1 spannland, 2 s. = 1 tunnland = $218\frac{3}{4}$ square stang, -	=	1.211 acres.
SWITZERLAND. — Legal, since 1823, for the Cantons of *Aarau, Basle, Berne, Freiburg, Lucerne, Solothurn, Vaud*; but not in general use:		
10 zoll = 1 fuss, 4 f. = 1 stab = 1 *aune* of France, - - - - -	=	1.3124 yards.
$2\frac{1}{2}$ stab = 1 toise *or* ruthe, - -	=	3.2809 "
Special and local —		
Basle: 12 zoll = 1 schuh *or* fuss, 10 s. = 1 ruthe,	=	3.33 "
$21\frac{1}{2}$ zoll = 1 braccio, - - -	=	0.5966 "
$44\frac{1}{2}$ " = 1 elle, - - - - -	=	1.2348 "
$46\frac{1}{2}$ " = 1 klafter.		
Berne: 12 zoll = 1 fuss, 6 f. = 1 klafter, $1\frac{2}{3}$ k. *or* 10 fuss = 1 ruthe, - - - -	=	9.6215 feet.
$1\frac{17}{20}$ fuss = 1 elle, - - - - -	=	0.5933 yard.
Geneva: 12 zoll = 1 pied, $5\frac{1}{3}$ p. = 1 toise, -	=	8.528 feet.
1 aune (*wholesale*), - - - -	=	1.299 yards.
1 aune (*retail*), - - - - -	=	1.25 "
Lausanne: $3\frac{3}{5}$ piede = 1 aune (*metrical, of France*), - - - - -	=	1.3123 "
9 piede = 1 ruthe.		
Neufchatel: 12 zoll = 1 fuss, 10 f. = 1 tois, -	=	9.621 feet.
$22\frac{3}{4}$ zoll = 1 elle, - - - -	=	0.608 yard.
St. Gall: 10 zoll = 1 fus, 4 f. = 1 stab, -	=	1.3123 "
Zurich: 12 zoll = 1 fus, 2 f. = 1 elle, - -	=	0.6561 "
5 elle = 1 ruthe.		
TRIPOLI (N. Africa): 8 robi = 1 pic, -	=	0.6041 "
1 pic *for ribbons,* - - - -	=	0.5285 "
TUNIS. — 8 robi = 1 pic, *for woollens,* -	=	0.736 "
1 pic, *for silks,* - - - - -	=	0.690 "
1 pic, *for linens,* - - - - -	=	0.5173 "

Foreign.		U. States.
TURKEY. — *Aleppo:* 8 robi = 1 pic,	=	0.7396 yard.
1 dra mesrour,	=	0.6089 "
1 dra stambouli,	=	0.7079 "
Bagdad: 1 guz,	=	0.8796 "
Bussorah: 1 guz = 2.6389 ft. 1 hadid,	=	0.9502 "
Constantinople: 1 halebi,	=	2.325 feet.
1 endrasi *or* archim,	=	2.255 "
1 pic stambouli,	=	0.7079 yard.
1 pic for silks,	=	0.7317 "
Damascus: 1 pic,	=	0.637 "
Smyrna: 1 indise,	=	0.6846 "
8 rob = 1 pic,	=	0.7302 "

West Indies.

In the islands of *Antigua, the Bahamas, Barbadoes, Barbuda, Dominica, Grenada, Jamaica, Les Saints, Montserrat, Nevis, Santa Cruz, St. John, St. Kitts', St. Thomas, St. Vincent, Tobago, Tortola,* the Measures of Length are the same as in Great Britain.

In *Deseade, Guadeloupe, Mariegalante, Martinique, St. Lucia:*
12 ponce = 1 pied de Roy, 3⅔ p. = 1 aune, = 1.3 yards. This being the old system of France, or system previous to 1812.

In *Bonaire, Saba, St. Eustatius, St. Martin,* Same as HOLLAND.

In *St. Bartholomew,* Same as SWEDEN.

In *Curacoa, Trinidad,* Same as CASTILE (*Spain*).

CUBA I. — *Cardenas, Cienfuegos, Havana, Matanzas, Nuevitas, Porto Principe, St. Jago,* &c.: Same as CASTILE (*Spain*).

HAYTI I. — *Aux Cayes, Cape Haytien, Jeremie, Port au Prince, Port Platte,* &c.: Same as France, before 1812.

Savanna, St. Dmingo, &c.: Same as CASTILE (*Spain*).

PORTO RICO I. — *Ponce, St. Johns,* &c.: Same as CASTILE (*Spain*).

FOREIGN WEIGHTS REDUCED TO UNITED STATES.

Foreign.	U. States. Avoirdupois pounds.
ABYSSINIA. — *Massuah:* 10 dirhem = 1 wakea, 12 w. *or* 10 mocha = 1 rotl *or* liter,	= 0.688
ALGIERS. — 8 mitkal = 1 wakea, 16 wakea = 1 rotl attari (*for spices and drugs*),	= 1.190
18 wakea = 1 rotl gheddari (*for fruits, oil,* &c.),	= 1.339
27 wakea = 1 rotl khebir (*market pound*),	= 2.008
100 rotl = 1 cantaro *or* quantar.	
1000 grammes = 1 kilogramme,	= 2.205
ARABIA. — *Hodeida:* 30 vakia = 1 maon, 10 maon = 1 frazil, 40 frazils = 1 bahar,	= 813.—
Jidda: 15 vakia = 1 rotolo, 5 rotoli = 1 maund,	= 1.83
10 maunds = 1 frazil, 10 f. = 1 bahar,	= 183.—
Mocha: 15 vakia *or* wakega = 1 rotolo *or* rotl,	= 1.5
2 rotolo = 1 maund *or* maon, 10 maunds = 1 frazil, 15 frazils = 1 bahar,	= 450.—
100 miscals = 1 vakia, 22½ v. = 1 maund copra,	= 2.25
At the Bazaar, *for coffee* —	
14¼ vakia = 1 rotolo, *and a* bahar,	= 435.—
Muscat: 24 cucha = 1 maund,	= 8.75
AUSTRIA. — *Trieste, Vienna:* 2 lothe = 1 unze, 4 u. = 1 vierding, 2 v. = 1 mark, 2 m. = 1 pfund, 20 p. = 1 stein, 5 s. = 1 centner,	= 123.47
4 centner = 1 karch. 1 saum,	= 339.5
1 saum *for steel,*	= 308.666
Trieste (*Venice weight*): 1 pfund (*peso grosso*),	= 1.052
1 pfund (*peso sottile*) — *apothecaries' weight,*	= 0.665
BOHEMIA. — 100 pfund = 1 centner,	= 113.4
Prague: 16 unze = 1 pfund,	= 1.26
18 pfund = 1 stein, 6 s. = 1 centner,	= 136.08
HUNGARY. — 1 oka,	= 2.78
AZORE ISLANDS. — Same as LISBON (*Portugal*).	
BALEARIC ISLANDS. — MAJORCA. — 25 rotoli = 1 arroba,	= 22.37
4 arroba = 1 quintal, 112 rotoli = 1 oder,	= 100.217
100 rotoli barbaresco = 1 quintal,	= 92.794

24 a   FOREIGN WEIGHTS REDUCED TO UNITED STATES.

Foreign.	U. States.
	Avoirdupois pounds.
MINORCA. — 100 libra = 1 cantaro,	= 88.2
100 rotoli barbaresco = 1 quintal,	= 81.727
3 quintals = 1 carga,	= 275.18
BELGIUM. — Same as HOLLAND.	
Previous to 1820 —	
8 pond = 1 stein, 12½ s. = 1 centner,	= 103.659
3 centner = 1 schippond, 4 centner = 1 charge.	
Waeg *of coals*,	= 150.—
BERMUDAS I. — Same as GREAT BRITAIN.	
BOURBON I. — 100 livres = 1 quintal, 3 q. = 1 charge,	= 323.765
BRAZIL. — 16 onça = 1 arratel, 32 arratel = 1 arroba,	= 32.501
4 arrobas = 1 quintal, 13½ q. = 1 tonelada.	

Central and South America.

Balize, Campeche, Guatimala, Honduras, Laguna, Leon, Nicaragua, San Juan, San Salvador, Sisal, &c.	
Buenos Ayres, Callao, Carthagena, Coquimbo, Guayaquil, Laguayra, Lima, Maracaybo, Montevideo, Rio Hacha, Truxillo, Valparaiso, &c. :	
16 onza = 1 libra, 25 l. = 1 arroba, 4 a. = 1 quintal,	= 101.546
Berbice, Demerara, Essequibo, Surinam. — Same as HOLLAND.	
Cayenne. — Same as FRANCE.	
CANARY ISLANDS. — Same as CASTILE (*Spain*).	
CANDIA I. — 44 oka = 1 cantaro,	= 116.568
CAPE COLONY. — *Cape Town.* — 32 loot = 1 pond,	= 1.03
CAPE VERDE ISLANDS. — Same as LISBON (*Portugal*).	
CHINA. — 10 lis = 1 tael *or* leung, 16 taels = 1 catty *or* kan, 100 catties = 1 pecul *or* tam,	= 133.333
22¾ chu = 1 leung, 16 l. = 1 catty, 2 c. = 1 yin,	
15 y. = 1 kwan, 3½ k. = 1 tam, 1⅕ t. = 1 shik,	= 160.—
CORSICA I. — 1 livre,	= 0.76
CYPRUS I. — 400 drachmi = 1 oka, 40 oka = 1 moosa,	= 112.—
1¼ oka = 1 rotolo,	= 5.25
DENMARK. — 32 lod = 16 unze = 2 mark = 1 pund,	= 1.101
100 pound = 1 centner,	= 110.11
12 pund = 1 bismerpund, 1½ b. = 1 lispund, 20 l. = 1 shippund,	= 352.364
2⅛ lispund = 1 waag,	= 39.639
EGYPT. — 144 drachmi *or* 100 miscali = 1 rotolo forforu,	= 0.95
400 drachmi = 1 harsela,	= 2.639
420 drachmi = 1 oka,	= 2.771

FOREIGN WEIGHTS REDUCED TO UNITED STATES. *a* 25

Foreign.	U. States.
	Avoirdupois pounds.

100 rotoli *or* 36 harseli = 1 cantaro forfora, — = 95.—
105 " *or* 36 oka = 1 cantaro, *for coffee*, - = 99.75
1 rotolo mina, *for spices*, - - - — = 1.403
1 " zaidino, *for dye-woods*, - - = 1.138
1 " zauro, *for iron*, - - - — = 2.216
102 rotoli = 1 cantaro, *for quicksilver and vermilion*, = 96.9
115 " = 1 " *for almonds and fruit*, - = 109.25
125 " = 1 " *for drugs*, - - — = 118.75
133 " = 1 " *for gum arabic*, - = 126.35
150 " = 1 " *for plumbago*, - — = 142.5

FRANCE. — 1000 milligrammes = 100 centigrammes =
   10 decigrammes = 1 gramme = 15.43315 troy
   grains.
100 grammes = 10 decagrammes = 1 hectogramme, = .0.22
10 hectogrammes = 1 kilogramme, - — = 2.205
10 kilogrammes = 1 myriagramme, - - = 22.047
100 myriagrammes = 10 quintal = 1 tonneau.
1 livre poids de marc, - - - — = 1.07922
1 livre metrique = ½ kilogramme, - - = 1.10237

GERMANY. — BADEN. — *Heidelberg, Manheim*, &c. :
1000 as = 100 pfennig = 10 centas = 1 zehnling.
1000 z. = 100 pfund = 10 stein = 1 centner, ● - = 110.237
8 quentchen = 2 loth = 1 unze.
16 unze = 2 mark = 1 pfund, - - = 1.102

BAVARIA. — 20 pfund (*legal*) = 1 stein, - — = 24.693
Augsburg: 512 pfennig = 128 quentchen = 32 loth
  = 16 unze = 2 mark = 1 pfund *or* frohngewicht, = 1.082
100 pfund = 1 centner, - - - — = 108.262
3 centner = 1 pfundschwer *or* schiffpfund, - = 324.786
7½ pfundschwer *or* 100 stein = 1 tonne.
22½ pfund = 1 stein, 5½ stein = 1 wage, - — = 129.914
Nuremberg: Denominations and relative value, same
  as *Augsburg*.
100 pfund = 1 centner, - - - = 112.432

HANOVER (*legal*) : 32 ortchen = 16 drachma *or* quentchen = 2 loth = 1 unze, 8 u. = 1 mark, 2 m. =
1 pfund, - - - - — = 1.079
14 pfund = 1 liespfund, - - - = 15.11
24 liespfund = 1 pfundschwer *or* last, - — = 362.648
20 pfund = 1 stein *for flax*, - - - = 21.586

Bremen: 32 ortchen = 8 quentchen = 2 loth = 1 unze,
8 u. = 1 mark, 2 m. = 1 pfund, - - = 1.099
116 pfund = 1 centner, - - - — = 127.516
2½ centner = 1 schiffpfund, - - - = 318.791

Foreign.   *U. States.*
Avoirdupois pounds.

```
300 pfund = 1 pfundschwer or last,        -  = 329.784
 14 pfund = 1 liespfund.  1 wage, for iron, - = 131.913
  1 stein, for flax = 21.985 lbs. 1 stein, for wool, = 10.993
```
Emden, Osnaburg: 16 unze = 2 mark = 1 pfund, 100
   pfund = 1 centner, - - - - = 109.041
   3 centner = 1 pfundschwer.

HESSE CASSEL. — 16 unze = 1 pfund, 100 pfund = 1
   centner, - - - - - = 106.762
   21 pfund = 1 kleuder.

HESSE DARMSTADT *(legal)*: 100 pfund = 1 centner, = 110.236
  Frankfort: 16 unze *or* 2 mark = 1 pfund, - = 1.114
   100 pfund = 1 centner, - - - = 111.407
   22 pfund = 1 stein, - - - - = 24.509

HOLSTEIN. — *Hamburg, Lubec, Altona, Kiel:*
   16 unze = 2 mark = 1 pfund, - - = 1.068
   14 pfund = 1 liespfund, 8 l. = 1 centner, - = 119.6
   2½ centner = 1 schiffpfund, - - = 299.—
   7¼ schiffpfund *or* 100 stein = 1 tonne, - = 2135.72
   90 pfund = 1 last, - - - - = 96.11
   1 stein, *for feathers and wool*, - - = 10 679
   320 pfunds *(land freight)* = 1 schiffpfund, = 341.72

MECKLENBURG *(generally)*: 16 unze = 2 mark = 1
   pfund, 15 pfund = 1 liespfund, 20 l. = 1 schiffpfund, - - - - - = 313.723
  Rostock: 16 pfund = 1 liespfund, 20 l. = 1 schiffpfund, = 358 54
   280 pfund = 1 schiffpfund, *for iron and lead*, - = 313.723
   22 pfund = 1 stein, *for wool and flax*, - = 24.65

SAXONY. — *Dresden, Leipsic:* 32 pfennig = 8 drachma *or*
   quentlein = 2 loth = 1 unze, 8 u. = 1 mark, 2
   mark = 1 pfund, - - - - = 1.03
   22 pfund = 1 stein, 2 s. = 1 wage, - - = 45.324
   110 pfund = 1 centner, - - - = 113.31
   114 " = 1 " berg-gewicht.
   118 " = 1 " stahl-gewicht.
  Freyberg: 1 pfund, - - - - = 1.165

OLDENBURG. — Denominations and relative values, same
   as *Bremen*, but weight values 2.94% less.

GREAT BRITAIN.* — See United States.

* In Great Britain, in addition to the denominations of weights used in the United States (the values of which are the same), the

Clove of wool, ... = 7 lbs.	Stone of butchers' meat or flesh, = 8 lbs.
Stone " " iron, flour, . = 14 "	Stone of cheese, ... = 16 "
Tod " " . . . . . . = 28 "	Stone of glass, ... = 5 "
Weigh " " . . . . . = 182 "	Seam of glass, ... = 120 "
Sack " " . . . . . = 364 "	Stone of hemp, ... = 32 "
Last " " . . . . . = 4368 "	Fother of lead, ... = 19½ cwt.

FOREIGN WEIGHTS REDUCED TO UNITED STATES.   *a* 27

Foreign.	U. States. Avoirdupois pounds.
GREECE. — *Athens:* 400 drachmi = 1 oka,	= 3.37
44 oka = 1 cantaro,	= 148.3
MOREA. — 1 cantaro, *generally*,	= 123.75
Patras: 400 drachmi = 1 oka,	= 2.643
44 oka = 1 cantaro *or* quintal,	= 116.3
HOLLAND (*legal*) : 10 korrel = 1 wigtje, 10 w. = 1 lood, 10 l. = 1 onz, 10 o. = 1 pond = 1 *kilogramme* of France,	= 2.204
2000 ponds = 1 vat (*shipping*),	=2204.74
Previous to 1820 —	
300 ponds = 1 schippond,	= 326.77
8 ponds = 1 steen,	= 8.71

India and Malaysia or East Indies.

AN-NAM. — *Cochin China* — *Saigon:* 16 luong = 1 can, 10 can = 1 yen, 5.y. = 1 binh,	= 68.876
2 binh = 1 ta, 5 ta = 1 quan,	= 688.76
Tonquin — *Kesho:* 100 catties = 1 pecul,	= 132.—
BIRMAH. — *Pegu:* 12½ tical = 1 abucco, 2 a. = 1 agito, 4 agiti = 1 vis,	= 3.393
33 tical = 1 catty, 3 c. = 1 vis,	= 3.393
Rangoon: 2 small rwés = 1 large rwé, 4 large rwés = 1 bai, 2 b. = 1 mu, 2 m. = 1 mat'h, 4 mat'hs = 1 kyat *or* ticul, 100 k. = 1 paitktha *or* vis,	= 3.65
BORNEO I. — 100 catty = 1 pecul,	= 135.633
CEYLON I. — *Colombo:* 500 pond = 1 bahar *or* candy,	= 500.—
CELEBES I. — *Macassar:* 100 catty = 1 pecul,	= 135.633
HINDOSTAN — *Bengal*, generally (*bazar weight*) : 10 mace = 1 khancha, 3 k. = 1 chattac, 10 c. = 1 dhurra *or* pussaree, 8 d. = 1 maon *or* maund,	= 82.133
Calcutta (*factory weight*) : 5 sicca = 1 chattac, 16 c. = 1 seer, 40 s. = 1 maund,	= 74.666
Bombay: 36 tanks *or* 15 pice = 1 tippree, 2 t. = 1 seer, 40 s. = 1 maund, 20 m. = 1 candy,	= 560.—
2½ tank-seers = 1 rupee-seer, for liquids = 1.54 lbs.	
Goa: 32 seers = 1 maund, 20 m. = 1 bahar,	= 495.—
Madras: 10 pagodas *or* varahuns = 1 pollam, 8 p. = 1 seer, 5 s. = 1 vis *or* visay, 8 v. = 1 maund *or* maon,	= 25.—
20 maunds = 1 candy *or* baruay.	= 500.—
Malabar Coast: 40 polams = 1 vis, 8 v. = 1 maon,	= 30.—
20 m. = 1 candy, 20 c. = 1 garce.	
Malabar (*interior*): 20 maon = 1 candy,	= 695.—

28a   FOREIGN WEIGHTS REDUCED TO UNITED STATES.

Foreign.   *U. States.*
                                                               Avoirdupois
                                                               pounds.

Mangalore: 6 sida = 1 vis, 8 v. = 1 maund, 20 maunds
  = 1 candy,   -   -   -   -   -   = 564.72
Massulipatam: 1½ nawtauk = 1 chittac, 6 c. = 1
  yabbolam, 2 y. = 1 puddalum, 2½ p. = 1 vis, 6⅜
  v. = maund, 20 m. = 1 candy, -.   -   -= 500.——
  21¼ vis = 1 pucca maund,   -   -   -= 80.——
Mysore, Seringapatam: 10 varahuns = 1 pollam, 40
  p. = 1 pussaree, 8 pussaree = 1 maon *or* maund, = 24.276
  20 maund = 1 bahar *or* candy.
Pondicherry: 10 varahuns = 1 poloin, 40 p. = 1 vis,
  8 v. = 1 maund, 20 m. = 1 candy,   -   -= 588 ——
Serampore, Tranquebar (legal): Same as DENMARK.
Sinde: 2 pice = 1 anna, 2 a. = 1 chittac, 16 c. = 1
  seer, 40 seers = 1 maund,   -   -   -= 74.666
Surat (new measure): 8 pice = 1 tippree, 2 t. = 1
  seer, 40 s. = 1 maund, 21 m. = 1 candy, -   = 300.——
  3 candi = 1 bhaur.
Tatta: 4 pice = 1 anna, 16 a. = 1 seer, 40 s. = 1
  maund,   -   -   -   -   -   -= 74.32
JAVA I. — *Batavia:* 16 tael = 1 catty, 1¼ c. = 1 goelak,
  66⅔ g. *or* 100 catty = 1 pecul, 3 p. = 1 bahar =
  16 m. 1 vis, 24 pollams of Madras,   -   -= 405.333
  4½ pecul = 1 great bahar.
  5 pecul = 1 timbang, *for grain*,   -   = 675.555
Bantam: 2¼ tael = 1 goelak, 100 g. = 1 catty, 2
  catty = 1 bahar, *for pepper*,   -   -= 405.333
MALACCA. — *Malacca:* 16 tael = 1 catty, 100 c. = 1 pecul, 3 peculs = 1 bahar,   -   -   -= 405.——
  2½ pinga = 1 tampang, 2 t. = 1 bedoor, 12 b. = 1
  hali, 1¼ h. = 1 kip, *for tin*,   -   -= 40.677
  2 buncals = 1 catty, *for gold and silver*,   -   = 2.049
PHILIPPINE I. — *Manilla,* &c.:
  22 piastres = 1 catty, 100 c. = 1 pecul,   -   = 139.443
  1 caban of rice (*usual*),   -   -   -= 133.——
  1 caban of cocoa,   -   -   -   -= 83.5
SIAM. — *Bangkok,* &c.: 4 tical = 1 tael, 2 t. = 1 catty,
  100 catty = 1 pecul,   -   -   -   -= 135.238
SINGAPORE I. — Same as MALACCA.
SOOLOO ISLANDS. — 10 mace = 1 tael, 16 t. = 1 catty,
  50 c. = 1 luchsa, 2 1. = 1 pecul,   -   -= 133.333
SUMATRA I. — 62 catties = 1 pecul,   -   -= 132.587
  24 tael = 1 salup, 2 s. = 1 ootan, 7½ o. = 1 nelli,
  for camphor, -   -   -   -   -= 29.333

FOREIGN WEIGHTS REDUCED TO UNITED STATES.   a 29

Foreign.	U. States.
	Avoirdupois pounds.
Acheen: 10 mace = 1 tael, 20 t. = 1 goolak, 1¼ g. = 1 catty, 36 c. = 1 maund,	76.986
5½ maund = 1 candil *or* bahar.	
Bencoolen: 46 catties = 1 maund,	98.371
5¾ maunds = 1 bahar,	557.416
IONIAN ISLANDS. — *Cephalonia, Corfu, Ithaca, Paxos, Zante,* &c.:	
Legal since 1817 : 100 libbra = 1 talento,	100.—
100 oke = 385 marcs.	
44 oke = 1 cantaro,	118.807
Cephalonia: 64 libbra = 1 barile, *for salt,*	67.262
Corfu: 100 libbra = 1 talento,	90.034
ITALY. — CARRARA. — 1 carrata, *for marble,* = 25 cubic palma = 12.764 cubic feet = 31⅕ centinajo *or* 29⅝ quintale of MODENA,	=2240.—
LOMBARDY AND VENICE.	
Government and Customs Measure:	
10 grani = 1 denaro, 10 d. = 1 grosso, 10 g. = 1 oncia, 10 o. = 1 libbra, 10 l. = 1 rubbio, 10 r. = 1 centinajo = 10 *myriagrammes* of France,	220.474
Special and local:	
Venice. — PESO grosso : 4 grani = 1 carato, 32 c. = 1 saggio, 6 s. = 1 oncia, 6 o. = 1 marco, 2 m. = 1 libbra,	1.052
25 libbre = 1 miro, 40 m. = 1 migliajo,	=1051.86
Peso sottile : 4 grani = 1 carato, 24 c. = 1 saggio, 6 s. = 1 oncia, 12 o. = 1 libbra,	0.666
100 libbre = 1 quintale, 4 q. = 1 carica,	266.332
NAPLES. — 20 acini = 1 trapeso, 30 t. = 1 oncia, 12 o. = 1 libbra, 26 l. = 1 rubbio,	18.387
150 libbra = 1 cantaro piccole,	106.08
33⅓ onci = 1 rotolo, 100 r. = 1 cantaro grosso,	196.45
SARDINIA. — *Genoa:* 24 grani = 1 denaro, 24 d. = 1 oncia, 12 o. = 1 libbra, 1¼ l. = 1 rotolo, 16¾ r. (25 libbre) = 1 rubbio, 4 r. = 1 centinajo, 1½ c. = 1 cantaro.	
Peso grosso : 1 centinajo,	76.863
Peso scarso : 1 centinajo,	69.875
Nice: 12 oncia = 1 libbra, 25 l. = 1 rubbio, 4 rubbi = 1 centinajo,	68.694
Turin: 25 libbre = 1 rubbio,	20.329
STATES OF THE CHURCH. — 1 libbra Italiana,	2.204
Ancona: 12 oncia = 1 lira, 100 l. = 1 cantaro,	72.942

C*

	U. States.
Foreign.	Avoidupois pounds.

Rome: 12 oncia = 1 libbra, 10 l. = 1 decina, 10 d. =
 1 cantaro, - - - - -= 74.763
 160 libbre = 1 cantaro. 250 libbre = 1 cantaro.
 1000 libbre = 1 migliajo, - - - = 747.633
TUSCANY. — *Leghorn, Florence, Pisa:*
 72 grani *or* 3 denari = 1 dramma, 96 dramme *or* 12
  oncia = 1 libbra, 100 l. = 1 centinajo, - -= 74.857
 10 centinaje = 1 migliajo.
 160 libbre = 1 cantaro *or* carara, *for wool, fish, &c.*, = 119.771
 50 rottoli = 1 cantaro generale (*old*), - = 112.29
JAPAN. — 160 rin = 10 pun = 1 its-go - - = 1.333 lbs.
 2,500 pun = 100 ischo = 10 itho = 1 its 'ko-koo = 333¼ lbs.
MALTA I. — 30 trapesi = 1 oncia, 30 o. = 1 rotl,
 100 rotl = 1 cantaro sottile, - - = 174.504
 110 rotl = 1 cantaro grosso.
MADEIRA I. — Denominations and relative values,
  same as LISBON, *Portugal:*
 32 arratel *or* libbra = 1 arroba, - - -= 32.349
MAURITIUS I. — *Port Louis:* 16 onces = 1 livre; = 1.08
MEXICO. — Standard same as CADIZ, *Spain.*
MOROCCO. — 100 rotl = 1 cantaro, - - = 118.723
 Tangiers: 1 rotl (*miners'*), - - - -= 1.06
 1 rotl (*market*), - - - - = 1.701
MOZAMBIQUE (*Africa*). — *Mozambique:* Same as *Portugal.*
NORWAY. — Same as DENMARK.
PERSIA. — *Bushire:* 3 cheki = 1 ratel, 7½ r. 1 maund
 tabree, 2 m. tabree = 1 maund show.
 1 maund show, bazar, - - - = 12.5
 1 " copra, " - - -= 7.3
 1 " show, factory, - - = 13.5
 1 " copra, " - - -= 7.75
 Tauris: 2 mascais = 1 dirhem, 50 d. = 1 ratel, 6
  ratel = 1 batman, - - - = 5.047
 Shiraz: 1 batman, - - - -= 10.125
PORTUGAL. — 576 grao *or* 24 escropulo *or* 8 outuava,
 = 1 onça, 16 o. = 4 quarta = 2 marca = 1 arratel, - - - - - = 1.016
 32 arratel = 1 arroba, 4 arrobe = 1 quintal, -= 130.06
 13½ quintale = 1 tonelada.
PRUSSIA (*legal since* 1820): 4 quentchen = 1 loth,
 2 l. = 1 unze, 8 u. = 1 mark, 2 m. 1 pfund, = 1.0312
 16½ pfund = 1 liespfund, 1½ l. = 1 stein, -= 22.687
 5 stein = 1 centner, 3 c. = 1 schiffpfund, - = 340.31

FOREIGN WEIGHTS REDUCED TO UNITED STATES. a 81

Foreign.	U. States.
	Avoirdupois pounds.

100 pfunds = 1 lagel, *for steel*, - - -= 103.116
1 Prussian mark = 1 Cologne mark.
Dantzic: 33 pfund = 1 stein, *for flax*, - - = 34.029
RUSSIA (*legal throughout the Empire*):
  3 zolotnik = 1 loth, 32 loth *or* 12 lana = 1 funt,
    40 funt = 1 pud, - - -= 36.067
  10 pud = 1 berkowitz, 3 b. = 1 paken, - =1082.02
  2 paken = 1 last.
Libau: 20 funt = 1 liespfunt, 20 l. = 1 schiffpfunt, = 364.168
  Hamburg weights are also used here.
Riga: 20 pfunde = 1 liespfund, 5 l. = 1 lof, -= 92.158
  4 lof = 1 berkowitz *or* schiffpfund, - - = 368.633
SARDINIA I. — *Cagliari*, &c.: 12 oncia = 1 libbra,
  26 l. = 1 rubbio, 4 r. = 1 cantarello, - -= 93.082
SICILY I. — *Messina*: 12 oncia = 1 libbra, = 0.707
  2½ libbre = 1 rotolo sottile, - -= 1.768
  33 oncia = 1 rotolo grosso, - - - = 1.945
  100 rotoli = 1 cantaro (*gross or net*).
Palermo: 250 libbre *or* 100 rotoli sottile = 1 cantaro
  sottile, - - - - - -= 175.04
  275 libbre *or* 100 rotoli grosso = 1 cantaro grosso, = 192.556
Syracuse: 250 libbre *or* 100 rotoli sottile = 1 cantaro
  sottile, - - - - - = 180.125
  275 libbre or 100 rotoli grosso = 1 cantaro grosso, = 198.137
SPAIN. — 16 Castilian onze = 1 Castilian libra (*Customs*), - - - - - -= 1.01546
Alicant: 12 onze = 1 libra menor (*minor*).
  18 onze = 1 libra mayor (*major*), - = 1.144
  24 l. mayor *or* 36 l. menor = 1 arroba, -= 27.456
  4 arrobe = 1 quintal, 2½ q. = 1 carga, - = 274.567
  8 carga = 1 tonelada.
  24 Castilian libre = 1 arroba, *for vermilion*, -= 24.871
  25 Castilian libre = 1 arroba of the Customs, - = 25.386
Barcelona: 25 libre = 1 arroba, - -= 22.14
Bilboa: 25 libre = 1 arroba, - - = 26.97
Cadiz (*Standard of* CASTILE): 8 onza = 1 marco,
  2 m. = 1 libra, 25 l. = 1 arroba, 4 a. =1 quintal, = 101.546
  20 quintale = tonelada.
Corunna, Ferrol: 16 onze = 1 libra sutil, 100 libre
  sutil = 1 quintal (*Castilian*), - -= 101.546
  20 onze = 1 libra gallega, 100 l. g. = 1 quintal, = 126.933
  25 libre = 1 arroba.
Gibraltar: 16 onze = 1 libra (*Castilian*), - = 1.01546
  16 ounces = 1 pound, 25 p. = 1 arroba, - -= 25.—

FOREIGN WEIGHTS REDUCED TO UNITED STATES.

Foreign.	U. States.
	Avoirdupois pounds.

Malaga.—Same as *Cadiz.*
1¼ quintal, *or* 3¼ barrile = 1 carga of raisins, -= 177.7
Santander: 100 libre = 1 quintal, - - = 152.28
Valencia: 12 onze = 1 libreta *or* libra menor, -= 0.784
   18 onze = 1 libra gruesa, - - - = 1.176
   35 libre menor = 1 arroba, 4 a. = 1 quintal,
   12½ arobe = 1 carga, - - - -= 338.413

SWEDEN.— *Stockholm,* &c. :
Viktualie-wigt *or* skal-wigt : 4 quintin = 1 lod, 2 l.
   = 1 untz, 16 u. = 1 skalpund, - - = 0.9375
   20 skalpund = 1 lispund, 20 l. = 1 skeppund, -= 375.—
   32 skalpund = 1 sten, - - - = 30.—
   12 skeppund = 1 last.
Metall-wigt *or* jern-wigt (*for iron, steel,* &c.) :
   20 mark = 1 markpund, 20 m. = 1 lispund,
   20 l. = 1 skeppund, - - - -= 300.—
   15 skeppund = 1 last.
   8¼ lispund = 1 waag *or* vog, *for tin,* - -= 123.75
Uppstadt-wigt (*inland weight*) :
   400 pund *or* 20 lispund = 1 skeppund, -= 315.674
Tachjern-wigt : 400 pund *or* 20 lisp. = 1 skeppund, = 453.47
Berg-wigt : 400 pund *or* 20 lisp. = 1 skeppund, = 348.822

SWITZERLAND (legal, since 1823, for the Cantons of
  *Aarau, Basle, Berne, Freiburg, Lucerne, Solothurn,
  Vaud;* but not in general use) :
8 gros = 1 unze, 8 u. = 1 mark, 2 m. = 1 livre *or*
  pfund = 1 *livre poids de marc* of France, = 1.07922
10 livres = 1 stein, 10 s. = 1 centner, - -= 107.922

Special and local:
Berne: 100 pfunde = 1 centner, - - = 114.9
Geneva: 24 grani = 1 denier, 24 d. = 1 once, 15 o.
   = 1 livre foible, - - - -= 1.0118
   18 once = 1 livre fort, - - - = 1.2141
Lausanne: 16 onces = 1 livre = 1 *livre poids de metrique*
  of France, - - - - -= 1.10237
Neufchatel: 8 onces = 1 marc, 2 m. = 1 livre, - = 1.14682
St. Gall: 10 unze = 1 loth, 2 l. = 1 pfund, -= 1.2914
Zurich: 18 unze = 1 pfund, - - - = 1.1637
Poids foible (*for silks,* &c.) : 2 lothe = 1 unze, 8 u.
   = 1 mark, 2 m. = 1 pfund, - - - = 1.0344

TRIPOLI.— (N. Africa) : 8 termini = 1 usano, 16 u.
   = 1 rotolo, 100 rotoli = 1 cantaro, - -= 111.214
   400 drachmi = 1 oke, - - - = 2.7429

FOREIGN WEIGHTS REDUCED TO UNITED STATES.   *a* 33

Foreign.   *U. States.*
                                                  Avoirdupois
                                                    pounds.
TUNIS. — 8 metical = 1 usano, 16 u. = 1 rotolo, 100
  rotoli = 1 cantaro, - - - = 109.155
TURKEY. — *Aleppo:* 266¾ meticals = 1 oke, - - = 2.81349
  480 meticals = 1 rotolo, 5 r. = 1 vesno, 20 v. = 1
  cantaro, - - - - - = 506.428
  3¼ rotoli = 1 batman, 10¼ b. = 1 cola, - - = 177.249
  30½ rotoli = 1 cantaro zurlo, - - - = 154.46
  400 meticals = 1 rotolo for Damasceno, - - = 4.22023
  453¼ meticals = 1 rotolo for Persian silks, = 4.78293
  466¾ meticals = 1 rotolo Tripolitan, - - = 4.92361
Bagdad: 2½ vakia = 1 oke, - - - = 2.74286
Bussorah: 100 miscals = 1 cheko, - - - = 1.02857
  24 vakia = 1 maund, - - - - = 116.——
  46 oke = 1 cuttra, - - - - = 136.482
  24 vakia attaree = 1 maund attaree, - - = 28.——
  76 vakia attaree = 1 maund sessee, - - = 88.6666
  4⅓ vakia attaree = 1 vakia.
Constantinople: 16 kara, kilot *or* taim = 1 dirhem,
  100 dirhem *or* 66⅔ meticals = 1 cheki *or* yusdrum,
  2 cheki = 1 rottel, 100 r. = 1 cantaro, - = 140.3
  13 7⁄11 rottel = 1 batman, 7½ b. = 1 cantaro.
  266⅔ meticals *or* 2 rottel = 1 oka.
  116⅔ meticals = 1 cheki, *for opium*, - - = 1.7578
Damascus: 400 meticals *or* 60 peso = 1 rotolo, 100
  rotoli = 1 cantaro, - - - - = 395.673
Smyrna: 100 drachmi *or* 66⅔ miscals = 1 cheko, 2¼
  cheki = 1 cequi, - - - - = 1.7578
  180 drachmi *or* 120 miscals = 1 rotolo, 13¼ r. = 1
  batman, 7½ b. *or* 100 rotoli = 1 cantaro, - = 126.571
  4 cheki = 1 oka, 45 o. = 1 cantaro.
  44 oke = 1 cantaro, *for tin*, &c., - - = 123.758

West Indies.

In the Islands of *Antigua, The Bahamas, Barbadoes,
  Barbuda, Dominica, Grenada, Jamaica, Les
  Saints, Montserrat, Nevis, St. Kitts, St. Vincent,
  Tobago, Tortola,* the commercial weights are the
  same as in Great Britain.
In *Deseade, Guadeloupe, Mariegalante, Martinique, St.
  Lucia:* 2 quartiers = 1 marc, 2 m. = 1 livre, 100
  l. = 1 quintal, - - - - = 107.922
  3 quintals = 1 charge, 3½ c. = 1 millier.
  This being the old system, poids de marc, of France.

Foreign.	U. States.
	Avoirdupois pounds.

In *Saba, St. Eustatia, St. Martin,* the commercial weights are the same as in Holland, *old system.*
In *Santa Cruz, St. John, St. Thomas,* Same as DENMARK.
In *St. Bartholomew,* Same as SWEDEN.
In *Curacoa, Trinidad,* Same as CASTILE (*Spain*).
In *Bonaire:* 100 pond = 1 centenaar,   -   -   - = 103.659
    3 centenaar = 1 schippond.
    This being the old weight of BRABANT, *Holland.*
CUBA I. — *Cardenas, Cienfuegos, Havana, Matanzas, Nuevitas, Porto Principe, St. Jago,* &c. : Same as CASTILE (*Spain*).
HAYTI I. — *Aux Cayes, Cape Haytien, Jeremie, Port au Prince, Port Platte,* &c. : Same as France. before 1812.
  Savanna, St. Domingo, &c. : Same as CASTILE (*Spain*).
PORTO RICO I. — *Ponce, St. Johns,* &c. : Same as CASTILE.

FOREIGN LIQUID MEASURES REDUCED TO UNITED STATES.

Foreign.	U. States Wine gallons
ABYSSINIA. — *Massuah:* 1 cuba, =	0.268
ALGIERS. — 16¾ litres = 1 khoulle, 6 k. = 1 hectolitre, =	26.418
ARABIA. — *Mocha:* 20 vacias (*weight*) = 1 nusfiah, 8 n. = 1 cuda *or* gudda = 16 lbs. Av., or *of oil*, &c. =	2.07
AUSTRIA (*legal and general for the Empire*):	
Vienna, Trieste, Lintz, Prague, Pesth, &c. :	
4 seidel = 1 mass, 10 mass = 1 viertel, =	3.738
4 viertel = 1 eimer *or* orna, =	14.952
32 eimer = 1 fuder, =	478.48
Special and local —	
Trieste: 4¾ caffiso = 1 orna, =	14.952
BOHEMIA. — 32 pinte = 1 eimer, =	16.141
4 eimer = 1 fass, =	64.56
Prague: 60 mass = 1 eimer, =	16.591
HUNGARY. = 4 rimpel = 1 halbe *or* icze, 100 icze = 1 czeber, =	22.—
1 anthal = 13.352 gals. 1 ako, =	18.495
Buda, Pesth, &c. : 1 eimer, =	15.028
Presburg: 1 eimer, =	19.368
MORAVIA. — 40 mass = 4 viertel = 1 eimer, =	11.282
AZORE ISLANDS. — Same as LISBON (*Portugal*).	
BALEARIC ISLES. — MAJORCA. — 4 quarta = 1 quartès, 6 quartès = 1 cuartin, 4 cuartin = 1 carga, =	28.666
3 carga = 1 pelexo. 1 quartinello, =	1.8
MINORCA. — 4 quarta = 1 quartès, 3 quartès = 1 gerrah, 10 g. = 1 carga, 4 c. = 1 botta, =	133.379
1 barrel, =	8.344
BELGIUM. — 3½ canette = 1 uper, 10 u. = 1 emmer, 3 e. = 1 vat = 1 *hectolitre* of France, =	26.418
2 pinte = 1 pot *or* mingle, 2 p. = 1 stoop *or* gelte, 2 s. = 1 schreef, 25 schreef = 1 aam, =	36.578
6¼ aam = 1 ton of spirits, for shipping, =	237.76
2¾ stoop = 1 velte, =	2.011

36 a FOREIGN LIQUID MEASURES REDUCED TO UNITED STATES.

Foreign.   *U. States.*
Wine gallons.

BERMUDAS ISLANDS. — Same as UNITED STATES.
BRAZIL. — *Bahia:* 10 garrafa = 1 canada,   -   = 18.734
   10 canada = 1 pipa of molasses,   -   -   = 187.342
   7¼ canada = 1 pipa of spirits, -   -   -   = 134.886
   Rio Janeiro: 4 quartilho = 1 medida, 3 m. = 1 pote, = 2.185
   16 pote = 1 pipa, 2. p. = 1 tonelada,   -   -   = 262.178
   1 frasco,   -   -   -   -   -   = 0.562

Central and South America.

BALIZE, GUATIMALA, YUCATAN, BOLIVIA, BUENOS AYRES, EQUADOR, NEW GRANADA, PARAGUAY, PERU, URUGUAY, VENEZUELA. Denominations and values, same as CASTILE (*Spain.*)

GUIANA. — *Berbice, Demerara, Essequibo, Surinam:* Same as HOLLAND.

Cayenne. — Same as FRANCE.

CANARY ISLANDS. — *Grand Canary, Teneriffe,* &c. :
   4 cuartilla = 1 arroba, 28¼ a. = 1 pipa,   -   = 120.06

CANDIA I. — *By weight* — 1 oke = 2.649 lbs. Av.

CAPE COLONY. — *Cape Town:* 16 flask = 1 anker,
   4 anker = 1 aam, 4 aam = 1 legger,   -   - = 152.—

CAPE VERDE I. — Same as LISBON (*Portugal*).

CHILI. — *Valparaiso, Coquimbo,* &c. : 4 copa = 1 quartilla, 4 q. = 1 arroba,   -   -   -   =. 9.906

CHINA. — By weight. See Weights. Also —
   10 kop tsong = 1 shing tsong, 10 shing tsong = 1 tau tsong, 5 tau tsong = 1 hok tsong, 2 hok tsong = 1 shik tsong, 1¾ shik tsong = 1 yu, 5 yu = 1 ping = 832 lbs. Av.

CORSICA I. — 4 cuarto = 1 boccale, 9 b. = 1 zucca, 12 z. = 1 barile,   -   -   -   - = 36.985

CYPRUS I. — By weight. See Weights.

DENMARK. — 155 pagel *or* 38¾ potte *or* 19⅜ kande = 1 anker,   -   -   -   -   -   = 9.889
   4 anker = 1 aam, 1½ a. = 1 oxehoved, 4 o. = 1 fuder, 1¼ f. = 1 stykfad,   -   - = 296.672
   60 viertel = 1 piba, -   -   -   -   = 122.499
   17 viertel = 1 toende, *for beer,*   -   - = 34.708

EGYPT. — By weight. See Weights.

FRANCE. — 1000 millilitres = 100 centilitres = 10 decilitres = 1 litre,   -   -   -   = 0.264
   100 litres = 10 decalitres = 1 hectolitre,   -   = ·26.418
   100 hectolitres = 10 kilolitres = 1 myrialitre.

FOREIGN LIQUID MEASURES REDUCED TO UNITED STATES. *a* 37

Foreign.	U. States. Wine gallons.

GERMANY. — BADEN (*legal*) : 4 schoppen = 1 mass,
   12½ m. = 1 stutz, 8 s. = 1 ohm = 1½ *hectolitres* of
   France, - - - - - = 39.026
   10 ohm = 1 fuder.
  Manheim: 16 schoppen *or* 4 eich-mass = 1 viertel, 12
   viertel = 1 ohm, - - - - = 25.285
  16 schoppen *or* 4 wirths-mass = 1 viertel, 24 viertel
   = 1 ohm = 16 *decalitres* of France, - - = 42.268
BAVARIA (*legal*) : 4 quartil = 1 mas *or* masskanne, 60
   masskanne = 1 eimer, - - - - = 16.944
  64   "   = 1 cimer, *for beer*, - - = 18.075
 Augsburg, Wurtzburg: 4 achtel = 1 seidel, 16 s. = 1
   beson, 8 beson = 1 cimer, 12 c. = 1 fuder, - = 238.745
  Nuremberg: visir-mass: 4 seidel = 1 viertel, 32 v.
   = 1 eimer, 12 c. = 1 fuder, - - - = 232.348
  Schenk-mass : 32 viertel = 1 eimer, - -= 18.244
  12 eimer = 1 fuder, - - - - = 218.924
  Ratisbon :. 32 viertel = 1 eimer, 12 e. = 1 fuder, -= 158.507
HANOVER (*legal*) : 8 nossel = 4 quartier = 2 kanne = 1
   stubchen, - - - - = 1.036
  10 stubchen = 5 viertel = 1 anker, - -= 10.36
  4 anker *or* 2½ cimer = 1 ahm, - - - = 41.439
  6 ahm *or* 4 oxhoft = 1 fuder, - - -= 248.637
  4 ahm *or* tonne, *for beer*, = 1 fass, - - = 165.756
  Bremen: 180 mingel = 90 versel = 45 quartier *or*
   . vierling = 11¼ stubchen = 5 viertel = 1 anker, = 9.574
  24 anker = 6 ohm = 1 fuder, - - = 229.779
  16 mingles = 1 stubchen, 6 s. = 1 stechkanne, 6
   stechkanne = 1 tonne, *for whale oil*, - -= 30.64
  44 stubchen (*beer measure*) = 1 tonne, - = 43.836
HESSE CASSEL : 4 schoppen = 1 mass, 4 m. = 1 viertel
   or quartlein, 20 v. = 1 ohm, 6 ohm = 1 fuder, = 251.547
  20 viertel (*beer measure*) = 1 ohm, - -= 46.128
HESSE DARMSTADT (*legal*) : Denominations and relative
   values, same as H. CASSEL.
  1 ohm = 16 *decalitres* of France, - - = 42.27
  Frankfort: 4 schoppen = 1 mass, 4½ neu-mass = 1
   viertel, 20 v. = 1 ohm, 6 ohm = 1 fuder, -= 227.352
  8 alt-mass = 9 neu-mass.
  1½ fuder = 1 stückfass, - - - = 303.136
HOLSTEIN. — *Hamburg, Altona, Lubec:* 8 oessel, plank *or*
   nossel = 4 quartier = 2 kanne = 1 stubchen, = 0.955
  8 stubchen = 4 viertel = 1 cimer, - - -= 7.644
  . 5 eimer *or* 4 anker = 1 ahm *or* fass, - = 38.22

38 a FOREIGN LIQUID MEASURES REDUCED TO UNITED STATES.

Foreign.	U. States.
	Wine gallons.
1½ ahm or 1¼ tonne = 1 oxhoft, - - - =	57.33
4 oxhoft or 2 pipe = 1 fuder, - - - =	229.32
16 margel or melgel = 1 stechkanne, 6 s. = 1 tonne,	
2 t. = 1 quarteel, *for whale oil*, - - - =	61.548
1 tonne, *for beer*, - - - - =	45.804

MECKLENBURG. — *Rostock*, &c. : Same as HOLSTEIN.

SAXONY. — *Dresden:* 8 quartier or 2 nossel = 1 kanne,
  quart or shenkkanne, 3 kanne = 1 viertel, - =   0.743
  18 v. = 1 anker, 1½ a. = 1 eimer, - - = 17.833
  2 e. = 1 anker; 1½ a. = 1 oxhoft, - - = 53.499
  1⅔ o. = 1 fass, 2⅖ f. = 1 fuder, - - = 213.996
  4 tonne or 2 viertel = 1 fass, *beer measure*, - = 104.026
 Leipsic: Denominations and relative values same as
  Dresden, but capacity values = 12.53 % greater.
  24 viertel = 1 eimer, - - - - = 20.067
 Freyburg: 100 schoppen or viertel = 1 brente, - = 10.312
  16 brente = 1 fass, - - - - = 165.—

GREAT BRITAIN : *Imperial measure:* Denominations
  and relative values same as *wine measure*, U. S.,
  but capacity values $20\frac{37}{115}$ per cent. greater. See
  LIQUID MEASURES, U. S.

GREECE. — *Patras:* 24 boccale = 1 barile, - - = 13.54

HOLLAND (*legal*) : 10 vingerhoed = 1 maatje, 10 m.
  = 1 kan, 10 k. = 1 vat = 1 *hectolitre* of F., = 26.418
 Previous to 1820 —
  2 mutsje = 1 pint, 2 p. = 1 mingle, 2 m. = 1 stoop,
  $3\frac{1}{27}$ s. = 1 viertel, 2⅛ v. 1 steekan, 2 s. = 1 anker,
  4 a. = 1 aam, 1½ aam = 1 okshoofd, - - = 61.—
  1 aam, *for oil*, - - - - = 37.64
  1 legger, *for beer*, - - - - = 153.57

India and Malaysia or East Indies.

CEYLON I. — *Colombo:* 4 aams = 1 legger, - - = 150.—

HINDOSTAN. — *Bombay:* 60 rupees = 1 seer, 50 seers =
  1 maund fluid = 77 lbs. Av.
 Calcutta: 5 sicca = 1 chattac, 4 c. = 1 pouah, 4 p.
  = 1 seer, 5 s. = 1 pussaree, 8 p. = 1 bazar
  maund weight = 82.13 lbs. Av.
 Madras: By weight. See Weights.
 Serampore, Tranquebar, (*legal*) : Same as DENMARK.

JAVA I. — *Batavia:* 5 kan = 1 barile, - - = 13.207

LUZON I. — *Manilla:* Same as CADIZ (*Spain*).

FOREIGN LIQUID MEASURES REDUCED TO UNITED STATES. a 39

Foreign.	U. States. Wine gallons.

SIAM. — *Bangkok:* 20 canan = 1 cohi = 5 *decalitres* of
France, - - - - - = 13.209
SUMATRA 1. — 8 pakha *or* 2 culah = 1 koolah, 15 koolah
= 1 tub, - - - - - = 17.44
IONIAN ISLANDS. — *Cephalonia:* 2 quartucci = 1
boccale, 8 b. = 1 pagliazza, 1¼ p. = 1 secchio, 6
s. = 1 barile, - - - - - = 18.—
Corfu, Paxos: 1¼ miltre = 1 boccale, 12 b. = 1 secchio, 6 s. = 1 barile, - - - = 18.—
Zante: 16 quartucci *or* 8 boccale = 1 lira, 1½ l. = 1
secchio, 6 s. = 1 barile, - - - - = 18.—
ITALY. — LOMBARDY AND VENICE. — *Government and Customs measure:* 10 coppi = 1 pinta, 10 p. = 1
mina, 10 m. = 1 soma = 1 *hectolitre* of France, = 26.418
Special and local —
Venice: 1¼ quartuzzi = 1 boccale, 2⅜ b. = 1 bozza, 4
bozzi = 1 secchio, 6 s. = 1 mastello *or* concia = 17.119
2 mastelli = 1 bigoncia, 4 b. = 1 anfora, - = 136.95
1¼ anfori = 1 botta, - - - - = 171.187
16 miri = 1 bigoncia, 2½ b. = 1 migliajo, 2 m. =
1 botta, *for oil*, - - - - = 322.22
NAPLES : 60 carraffa = 1 barile, - - - = 11.581
3½ barile = 1 salma, 4 s. = 1 pipa, - = 162.137
12 barile = 1 botta, 2 botte = 1 carro.
6 misurella *or* 1¼ pignata = 1 quarto, 16 q. = 1
stajo, 16 s. = 1 salma, *for oil*, - - = 42.538
11 salma = 1 last for shipping.
SARDINIA. — *Genoa:* 90 amola *or* 5 foglietta *or* pinta =
1 barile, 2 b. = 1 mezzaruola, - - - = 39.218
16 quarteroni = 1 quarto, 4 q. = 1 barile, *for oil*, = 17.084
Turin: 20 quartine = 1 boccale, 2 b. = 1 pinta, 6 p.
= 1 rubbio, 6 r. = 1 brenta, 10 b. = 1 carro, = 148.806
STATES OF THE CHURCH. — *Ancona:*
4 fogliette = 1 boccale, 24 b. = 1 barile, - = 11.35
1 soma of oil, - - - - - = 18.494
Rome: Wine measure: 4 cartocci = 1 quartuccio, 4
q. = 1 foglietta, 4 f. = 1 boccale, - - = 0.481
32 boccali = 1 barile, 16 b. = 1 botta, - = 246.544
Oil measure: 4 boccali = 1 cugnatello, 10 c. = 1
mastello *or* pello, 2 m. = 1 soma, - - = 43.333
TUSCANY. — *Leghorn, Florence, Pisa:*
4 quartucci *or* 2 mezette = 1 boccale, 40 b. *or* 20
fiasco = 1 barile = 133½ libbra *or* 99.81 lbs. Av.

40 a FOREIGN LIQUID MEASURES REDUCED TO UNITED STATES.

	Foreign.	U. States Wine gallons

    16 fiasco = 1 barile, 2 b. = 1 botta, *for oil* = 240 libbra *or* 179⅔ lbs. U. S.
MALTA I. — 2 caffisi = 1 barile = 50 rotl *or* 87¼ lbs. Av.
MADEIRA I. — Standard same as LISBON (*Portugal*).
MAURITIUS I. — *Port Louis:* 1 velt. - - = 2.—
MEXICO. — Same as CADIZ (*Spain*).
NORWAY. — Same as DENMARK.
PORTUGAL. — *Lisbon*, &c.: 24 quartilhi *or* 6 canada
  = 1 alqueire *or* cantaro, - - - -= 2.185
  2 alqueire = 1 almude, 26 a. = 1 bota *or* pipa, = 113.627
  2 bota = 1 tonelada.
  31 almudes = 1 pipa, London gauge.
Oporto: 2 alqueire = 1 almude, 21 a. = 1 pipa, = 139.134
PRUSSIA (*legal throughout the kingdom since* 1820 :)
  2 ossel = 1 quart, 30 q. = 1 anker, 2 a. = 1 eimer
  = 3840 cubic Rhein-zolle *or* 4192 cubic inches,
  U. S., - - - - = 18.146
  3 eimer *or* 1½ ohm = 1 oxhoft, 4 o. = 1 fuder, -= 217.758
  3¼ eimer = 1 fass.
Dantzic: 3¼ eimer = 1 fass, - - - = 60.487
  6 cimer = 1 pipe, - - - -= 108.876
Konigsberg: 4½ cimer = 1 pipe, - - = 81.658
RUSSIA (*legal for the Empire since* 1820) :
  12¼ tscharka = 1 osmuschka *or* krashka, 2 o. = 1
    tschet-werk, 4 t. = 1 vedro *or* wedro, - -= 3.246
  40 vedro = 1 botschka *or* anker, - - =: 129.86
  13¼ botschka = 1 sarokowaja.
Revel, Riga: 30 stof = 5 viertel = 1 anker, - -= 10.311
  4 anker = 1 ahm, 6 a. = 1 fuder, - - = 247.46
SARDINIA I. — 4 quartucci = 1 quartaro, 8 q. = 1
  barile, - - - - - -= 8.874
  2 barile = 1 mezzaruola.
SICILY I. — *Palermo:* 20 quartucci = 1 quartaro, 8
  quartari = 1 barile, - - - - = 9.436
  12 barile *or* 5 salma = 1 botta *or* pipa, - -= 113.237
  12 salma = 1 tonnellata.
  1 caffiso, *for oil* = 12½ rotoli grosso, - - = 3.09
Messina: 12 barile *or* 5 salma = 1 pipa, - -= 108.—
  1 caffiso, *for oil* = 12½ rotoli grosso.
Syracuse: 12 salma = 1 tonnellata, - - -= 247.—
SPAIN. — For Customs values, see *Cadiz*.
Alicant: 4 copa = 1 cuartilla, 4 c. = 1 cuarto, 4
  cuarti = 1 cantaro = 1 Castilian arroba = 25.38

FOREIGN LIQUID MEASURES REDUCED TO UNITED STATES. *a* 41

Foreign. *U. States.*
Wine gallons.

lbs. = 3.04 gallons of wine or 3.642 gallons of 90 per cent. alcohol, United States measure.
40 arrobe = 1 pipa, 2 p. = 1 tonelada.
Barcelona: 4 petricon = 1 mitadella *or* porrone, 4 m. = 1 quartera, 2 q. = 1 cortan *or* mitjera, 2 c. = 1 mallah, 8 m. = 1 carga = 12 arrobe *or* 265.68 lbs. Av.
4 carga = 1 pipa.
4 quarta = 1 cuarto, 4 c. = 1 cortan, 30 cortan = 1 carga, *for oil* = 11 arrobe or 243.54 lbs. Av.
Cadiz (*Standard of* Castile): 2 copa = 1 azumbra, 2 a. = 1 cuartilla, 4 c. = 1 cantaro *or* arroba.
1 arroba mayor = 35 libre or 35.541 lbs. Av., or 984¾ cubic inches of distilled water at 60°, — = 4.263
1 arroba menor, *for oil* = 27¼ libre, - - = 3.319
16 arrobe = 1 mayo, 27 arrobe = 1 pipa, 30 arrobe = 1 bota, 2 bote = 1 tonelada.
Corunna, Ferrol: 16 quartilli = 1 olla, 4 o. = 1 canado, 4 c. = 1 mayo = 14 arroba sutil, - - = 42.533
Gibraltar: Same as the United States.
Malaga: 8 azumbre = 1 cantaro *or* arroba (34½ l.), = 4.182
Santander: 8 azumbre = 1 cantaro = 26 libre, - = 4.739
Valencia: 1 arroba = 38 libre menor, - - = 3.566
1 arroba, *for oil* = 29¼ libre menor, - - = 2.737
SWEDEN.— *Stockholm,* &c.: 4 jungfru *or* ort = 1 quarter, 4 q. = 1 stop, 2 s. = 1 kanna, - -= 0.689
6 kanne = 1 utting *or* ottingar, 2 a. = 1 fjerding,
1¼ f. = 1 ankare, 2 a. = 1 embar, 1⅝ e. = 1 tunna, - - - - - - = 33.174
1¼ tunna = 1 am, 1½ a. = 1 oxhufwud, - -= 62.202
2 oxhufwud = 1 pipa, 2 p. = 1 fuder, - = 248.81
SWITZERLAND (legal since 1823, for the Cantons of *Aarau, Basle, Berne, Freiburg, Lucerne, Solothurn, Vaud;* but not in general use):
10 emine = 1 mass *or* pot, 10 m. = 1 gelt = 1⁷⁄₁₀ decalitres of Franco, - - - -= 3.5664
Special and local —
Basle: 4 schoppen = 1 mass, 4 m. = 1 viertel, 2⅔ v. = 1 setier, 1⅜ s. = 1 ohm, 3 o. = 1 saum, - = 40 337
Berne: 8 becher *or* 2 viertel = 1 mass, 25 m. = 1 brente *or* eimer, 4 b. = 1 saum, - - -= 44.161
4 saum = 1 fass, 1½ f. = 1 landfass, - - = 264.966
Geneva: 2 pot = 1 quarteron, 24 q. = 1 setier, -= 11.942

D*

42a FOREIGN LIQUID MEASURES REDUCED TO UNITED STATES.

Foreign.   U. States.
                                                         Wine
                                                         gallons.

12 setier = 1 char.
Lausanne: 10 verre = 1 mass, 10 m. = 1 broc, 3 b.
    = 1 setier or eimer, - - - -= 10.699
Neufchatel: 8 pote or mass = 1 brochet or stutz,   = 4.024
2½ brochet = 1 brande, 2⅗ brande = 1 gerl, -= 26.159
24 brochet = 1 muid, 2¼ m. = 1 bosse.
St. Gall: lauter-mass: 8 mass = 1 viertel, -   = 2.773
4 viertel = 1 eimer, 4 e. = 1 saum, - -= 44.37
7½ saum = 1 fuder.
Schenk-mass: 8 mass = 1 viertel, - - = 2.465
4 viertel = 1 cimer, &c.
8 mass = 1 viertel, for oil, - - -= 2.867
Zurich: lauter-mass: 8 statz or 2 mass = 1 kopf, = 0.9637
7½ kopf = 1 viertel, 4 v. = 1 eimer, 1½ e. = 1 saum, = 43.368
Schenk-mass: 2 quartli = 1 mass, 2 m. = 1 kopf, = 0.86746
7½ kopf = 1 viertel, &c.
TRIPOLI (N. Africa): 14 caraffa = 1 mataro, for oil
    = 42 rotoli = 46 $\frac{7}{10}$ lbs. Av.
1 barril = 116⅔ rotoli.
TUNIS. — 2 mettar, for wine = 1 mettar, for oil = 36
    rotoli = 39 $\frac{3}{10}$ lbs. Av.
1 millerolle = 120 rotoli.
TURKEY. — 1 almud, - - - - = 1.38

West Indies.

In the islands of *Antigua, the Bahamas, Barbadoes, Barbuda, Dominica, Grenada, Les Saints, Montserrat, Nevis, St. Kitts', St. Vincent, Tobago, Tortola,* the Measures for Liquids are the same as those of the United States, or the same as those of Great Britain, previous to 1825.
In *Jamaica*: 85 Imperial gallons = 1 puncheon, - = 102.03
In the Islands of *Deseade, Guadeloupe, Mariegalante, Martinique, St. Lucia*: 8 muces = 4 roquilles = 2 chopines = 1 pinte, 8 pintes = 4 pots = 2 gallons
    = 1 velt, - - - - - -= 2.—
35 veltes = 1 muid.
This was the system for Liquid Measures in France, before 1812, except that the value of the *muid* was 70.855 gallons, U. S.
In *Bonaire, Curacoa, Saba, St. Eustatius, St. Martin:* Same as Holland *old* measure, or measure before 1820.

Foreign.	U. States.
	Wine gallons

In *St. Bartholomew:* Same as SWEDEN.
In *Trinidad:* Same as CASTILE (*Spain*).
In *Santa Cruz, St. John, St. Thomas:* Same as DENMARK.
CUBA I. — *Cardenas, Cienfuegos, Matanzas, Nuevitas, Porto Principe, St. Jago,* &c.: Same as CASTILE (*Spain*).
 Havana: 1 arroba = 4.1 gallons. 1 bocoy, - = 36.—
HAYTI I. — *Aux Cayes, Cape Haytien, Jeremie, Port au Prince, Port Platte,* &c.: Same as France before 1812.
 Savanna, St. Dmingo, &c.: Same as CASTILE.
PORTO RICO I. — Same as CASTILE.

FOREIGN DRY MEASURES REDUCED TO UNITED STATES.

Foreign.	U. States. Winchester bushels.
ABYSSINIA. — *Massuah:* 24 madega = 1 ardeb, =	0.333
ALGIERS. — 2 tarrie = 1 saa *or* saha, =	1.125
8 saa = 1 caffiso, =	9.—
100 litres = 1 hectolitre, =	2.838
ARABIA. — *Mocha:* 40 kella = 1 tomaun (*for rice*) = 168 lbs. Av.	
AUSTRIA (*legal and general*):	
2 becher = 1 fudermassel, 2 f. = 1 muhlmassel, 2 m. = 1 achtel, 2 a. = 1 viertel, 4 v. = 1 metze, 30 m. = 1 muth, =	52.354
Local and special —	
Trieste: 2 polonic = 1 metze, 1¼ m. = 1 stajo, =	2.156
BOHEMIA. — *Prague:* 12 seidel = 1 massel, 4 m. = 1 viertel, 4 v. = 1 strich, =	2.656
3 viertel = 1 metze, =	1.992
HUNGARY: 32 halbe = 1 viertel, 4 v. = 1 metze, =	1.745
Buda and *Pesth:* 1 metze, =	2.27
MORAVIA: 1 metze, =	2.—
AZORE I. — 2 meio = 1 alqueire, 4 a. = 1 fanga, =	1.359
BALEARIC I. — MAJORCA: 6 barcella = 1 quartera, =	2.042
MINORCA: 6 barcella = 1 quartera, =	2.156
BELGIUM. — 100 uper = 10 setier = 3 mudde = 1 muid = 3 *hectolitres* of France, =	8.513
10 muid = 1 last, =	85.134
Antwerp, Brussels, &c. (*old measures*) —	
10 malster vat = 1 halster, 10 h. = 1 sac, =	6.918
108 gelte = 1 muid, =	8.302
Ghent: 1 halster, =	1.499
4 meuke = 1 raziere, =	2.26
Mechlin: 1 meuko, =	0.614
BERMUDAS I. — Same as UNITED STATES.	
BRAZIL. — 16 quarta = 1 fanga, 15 f. = 1 moio, =	23.02
Bahia: 1 alqueire, =	0.863
Rio Janeiro: 1 alqueire, =	1.135

FOREIGN DRY MEASURES REDUCED TO UNITED STATES.   a 45

Foreign.   U. States.
                                                       Winchester
                                                        bushels.

Central and South America.

Balize, Campeche, Nicaragua, San Salvador, Sisal, &c.
Buenos Ayres, Callao, Carthagena, Laguayra, Mara-
   caybo, Montevideo, Truxillo, Valparaiso, &c. : Same
   as CADIZ, generally.
Buenos Ayres : 1 fanega,   -   -   -   =   3.752
Montevideo : 1 fanega,   -   -   -   =   3.868
Valparaiso : 1 fanega,   -   -   -   =   2.572
Berbice, Demerara, Essequibo, Surinam : Same as HOL-
   LAND before 1820.
Cayenne : Same as FRANCE.
CANARY ISLANDS. — 12 celemine = 1 fanega,   =   1.776
   17 celemine = 1 fanega (*heaped*).
CANADA EAST. — 1 minot,   -   -   -   =   1.111
CANDIA I. — 1 carga,   -   -   -   =   4.323
CAPE COLONY. — *Cape Town :* 4 schepel = 1 muid,
   10 muid = 1 load,   -   -   -   =   30.65
CHINA. — By weight. See Weights.
CORSICA I. — 6 bacino = 1 mezzino, 2 m. = 1 stajo, = 4.256
CYPRESS I. — By weight. See Weights.
DENMARK. — 4 sextingkar *or* fjerdingkar = 1 otting-
   kar *or* skieppe, 2 o. = 1 fjerding *or* stubchen, 4
   f. = 1 toende,   -   -   -   =   3.947
   22 toende = 1 last,   -   -   -   =   86.836
EGYPT. — *Alexandria, Rosetta :* 1 kisloz,   -   =   4.85
   24 robi = 1 rebeb,   -   -   -   =   4.462
   Cairo : 24 robi = 1 ardeb,   -   -   =   5.165
FRANCE. — 100 litres = 10 decalitres = 1 hectolitre, = 2.838
   100 hectolitres = 10 kilolitres = 1 myrialitre, - = 283.782
GERMANY. — BADEN (*legal*) : 1000 becher *or* 100 massel *or* masslein *or* 10 sester = 1 malter,   - =   4.256
   10 malter = 1 zober = 15 *hectolitres* of France, - = 42.567
   Manheim, Heidelberg : 32 masschel *or* 4 immel *or* invel
   or 2 kumpf *or* vierling = 1 simmer, 2 s. = 1 vi-
   ernzel, 8 v. = 1 malter, *for wheat*, -   -   =   3.152
   1 malter, *minim*,   -   -   -   =   2.922
   1   "   for barley and oats,   -   -   =   3.546
BAVARIA (*legal*) : 8 masslein *or* 4 dreissiger = 1 achtel
   or massel, 4 achtel = 1 viertel, -   -   =   0.526
   12 viertel = 1 scheffel, -   -   -   =   6.31
   144 metzen *or* 12 mass = 1 scheffel, *for oats, &c.*, = 7.363
   4 kubel = 1 seidel, 6 s. *or* 4 scheffel = 1 muth, *for
   coals and lime*,   -   -   -   =   25.24

46 a FOREIGN DRY MEASURES REDUCED TO UNITED STATES.

Foreign.	U. States. Winchester bushels.
Bamberg: 40 gaissil = 1 simra, 3 s. = 1 sheffel, =	6.618
Bayreuth: 16 mass = 1 simmer, - - - =	14.044
Nuremberg: 16 mass *or* 2 diethaufe = 1 metze, 16 metzen = 1 malter *or* simmer, - - - =	9.028
16 hafer-mass = 1 hafer-metze, 32 hafer-metzen = 1 hafer-simmer, - - - - =	16.696
Ratisbon: 4 massel = 1 strich, - - - =	0.756
1 strich, *for salt,* &c., - - - =	1.51
Wurzberg: 144 massel = 12 mass = 2 achtel = 1 scheffel, - - - - - =	5.183
1 scheffel, *for oats,* &c., - - - =	8.533
HANOVER (*legal*): 144 krus = 24 vierfas = 18 drittel *or* metzen = 6 himten = 1 malter, - - =	5.296
8 malter = 1 wispel, 2 w. = 1 last, - - =	84.736
12 malter = 1 fuder, - - - - =	63.552
Bremen: 16 spint = 4 viertel = 1 scheffel, - - =	2.021
40 scheffel = 1 last, - - - - =	80.834
HESSE CASSEL: 64 kopfchen = 32 masschen = 8 metzen = 4 mass = 2 himten = 1 scheffel, - - =	2.28
3 scheffel *or* 1½ viertel *or* butte = 1 malter, - =	6.482
HESSE DARMSTADT (*legal*): 64 kopfchen = 32 masschen = 8 gescheid = 2 kumpf = 1 metze, - - =	0.452
2 metzen = 1 simmer, 4 s. = 1 malter, - - =	3.632
4 butte (*coal measure*) = 1 mass, - - - =	17.736
Frankfort: 4 schrott = 1 mass, 4 m. = 1 gescheid, 4 g. = 1 sechter, 2 s. = 1 metze, 2 m. = 1 simmer, 4 s. = 1 achtel *or* malter, - - - =	3.256
HOLSTEIN. — *Hamburg, Altona:* 4 masschen = 1 spint, 4 s. = 1 himt, 2 h. = 1 fass, - - =	1.495
20 fass = 10 scheffel = 1 wispel, - - =	29.892
4½ wispel *or* 1½ last = 1 stock, - - =	134.514
10 scheffel (*for barley and oats*) = 1 wispel, - =	44.831
45 tonne (*for coals*) or 30 sacks = 1 fass, - =	179.6
Lubec: 4 fass = 1 scheffel, 4 s. = 1 tonne, - =	3.796
3 tonne = 1 dromt, 8 d. = 1 last, - - =	91.1
96 scheffel or 24 tonne = 1 last, *for oats,* - =	106.918
Kiel: 4 scheffel = 1 tonne *or* barril, - - =	3.367
MECKLENBURG. — *Rostock,* &c.: 16 spint = 4 fass = 1 scheffel, 2 s. = 1 dromt, - - =	13.243
2⅔ dromt = 1 wispel, 3 w. = 1 last, - - =	105.944
45 viertel *or* 15 stubchen = 1 dromt, *for oats,* - =	14.9
SAXONY. — *Dresden, Leipsic:* 4 masschen = 1 metze, 4 m. = 1 viertel, 4 v. = 1 scheffel, - - =	2.963
12 s. = 1 malter, 2 m. = 1 wispel, - - =	71.11

FOREIGN DRY MEASURES REDUCED TO UNITED STATES. a 47

Foreign.	U. States. Winchester bushels.

GREAT BRITAIN: *Imperial measure:* See DRY MEAS-
URES of the United States; 1 bushel, — = 1.0315
GREECE. — *Patras:* 2 medimni = 1 staro, = 2.23
  1 bachel, — — = 0.85
HOLLAND (*legal*): 10 maatje = 1 kop, 10 k. = 1
  scheppel, 10 s. = 1 mudde *or* zac, 30 m. = 1 last
  = 3 *kilolitres* of France, — — — = 85.134

India and Malaysia or East Indies.

BIRMAH. — *Rangoon:* 2 lamyet = 1 lamé, 2 l. = 1 salé,
  4 s. = 1 pyis, 2 p. = 1 sarot, 2 s. = 1 sait, 4
  sait = 1 ten *or* basket = 16 vis = 58.4 lbs. Av.
CEYLON I. — *Colombo:* 24 seers = 1 parah, — — = 0.721
HINDOSTAN. — *Bombay: Salt measure* —
  10½ adowlies = 1 parah, 100 p. = 1 anna = 93.033
  cub. feet, 16 anna = 1 rash, — — =1196.13
  Grain measure — 2 tipprees = 1 seer, 4 s. = 1 adoulie,
  16 a. *or* 7 pallic = 1 para = 186⅔ lbs. Av.
  8 para = 1 candy.
  Calcutta: 5 chattac = 1 khoonka, 16 k. = 1 raik, 4 r.
  = 1 pallic, 20 p. = 1 soallic = 154 lbs. Av.
  8 soallic = 1 morah *or* maund *bazar.*
  16 morah = 1 kahoon.
  1⅛ bazar maunds = 1 soallic.
  Madras: 8 ollock = 1 puddy, 8 p. = 1 marcal, 5 m.
  = 1 para, — — — — — = 1.744
  80 para = 1 garce, — — — — = 139.535
  Serampore, Tranquebar (*legal*): Same as DENMARK.
  Tatta: 4 puttoes = 1 twier, 4 t. = 1 cossa, 60 c. =
  1 carvel = 5½ Tatta maunds *or* 408¾ lbs. Av.
JAVA I. — *Batavia:* 22 mudden = 1 coyang, — = 62.432
  Bantam: 1600 bambou = 400 gantang = 52 mudde
  = 1 coyang, *for rice,* — — — = 147.565
LUZON I. — *Manilla:* Same as CADIZ (*Spain*).
MALACCA. — 32 mudde = 1 coyang, — — — = 90.81
SIAM. — 40 sat = 1 sesti, 40 s. = 1 cohi, 10 c. = 1 co-
  yang = 32 *hectolitres* of France, — — = 90.81
SUMATRA I. — 4 pakha = 1 culah, 2 c. = 1 koolah, 15
  koolah = 1 tub, — — — — = 1.872
IONIAN ISLANDS. — *Corfu, Paxos:* 2 misura = 1
  bacile, 4 b. = 1 moggio, — — — = 4.777
  Cephalonia: 8 misure = 4 bacile = 1 moggio, — = 5.6
  Zante: 8 misure *or* 4 bacile = 1 moggio, — = 5.116

48 a FOREIGN DRY MEASURES REDUCED TO UNITED STATES.

	Foreign.	U. States. Winchester bushels.
Ithaca:	5 bacile = 1 moggio,	= 5.—

ITALY.—LOMBARDY AND VENICE.— *Government and Customs Measure:* 10 coppi = 1 pinta, 10 p. = 1 mina, 10 m. = 1 soma = 1 *hectolitre* of France, = 2.838

Special and local —
Venice: 4 quartaroli = 1 quarto, 4 q. = 1 stajo or staro, 4 s. = 1 moggio, - - - = 9.08

NAPLES. — 3 misura = 1 stopello, 4 s. = 1 mezetta, 2 m. = 1 tomolo, 36 t. = 1 carro, - - - = 54.81

SARDINIA. — *Genoa:* 12 gombetti = 1 ottaro *or* quarto, 8 quarto = 1 mina, - - - = 3.426
8 mine = 1 mondino, *for salt.*

Nice: 4 motureau = 1 quartier, 4 q. = 1 stajo, 4 staji = 1 sacco, - - - = 3.405

Turin: 20 cucchiari = 1 copello, 4 c. = 1 quartiere, 2 q. = 1 mina, 2 m. = 1 stajo, 3 s. = 1 sacco, = 3.263

STATES OF THE CHURCH.— *Ancona:*
4 provenda = 1 coppa *or* lappa, 2 c. = 1 corba, = 2.03

Rome: 5½ quartucci *or* 1½ scorzi = 1 starello, 2 starelli *or* 1½ staji = 1 quartarello, - - - = 1.044
4 quartarelli *or* 2 quarto = 1 rubbiatilla, 2 rubbiatille = 1 rubbio, - - - = 8.356
4¼ rubbi = 1 tonnellata (*shipping*).

TUSCANY. — *Florence, Leghorn, Pisa:*
8 bussole = 4 quartucci = 2 mezette = 1 metadella, 4 m. = 1 quarto, 4 q. *or* 2 mina = 1 stajo, 3 s. = 1 sacco, 8 sacci = 1 maggio, - - - = 16.592

JAPAN.— 10 gantang = 1 ickoga, 100 ickoga = 1 icmagoga, 100 icmagoga = 1 managoga.

MALTA. — 1 salma (rasa), - - - = 8.22
1 salma (colma), - - - = 9.56

MADEIRA I. — Standard same as LISBON (*Portugal*).
MEXICO. — Same as CADIZ (*Spain*).
MOROCCO. — *Mogador:* 1 mud, - - - = 5.184
NORWAY. — Same as DENMARK.
PORTUGAL. — *Lisbon, St. Ubes,* &c. :
16 quarto = 8 ineio = 4 alqueire = 1 fanga, - = 1.534
15 fanga = 1 majo, 4 m. = 1 last, - - = 92.087
27 fanga = 1 tonelada, *for shipping.*
1 balde, *for coals,* - - - = 12.69
1 fanga, " " = 21.167
Oporto: 1 fanga = 1.937 bus. 1 raze, *for salt,* = 1.25
PRUSSIA (*legal since* 1820) : 4 masschen = 1 metze, 4

FOREIGN DRY MEASURES REDUCED TO UNITED STATES. a 49

Foreign.	U. States. Winchester bushels.

metze = 1 viertel, 4 v. = 1 scheffel = 3072 cubic
Rhein zolle *or* 3353¾ cubic inches, U. S., -   -= 1.559
12 scheffel = 1 malter *or* dromt, 2 m. = 1 wispel, = 37.431
3 wispel = 1 last.
2 wispel = 1 last, for barley and oats.
RUSSIA (*legal for the Empire*) :
  8 garnetz = 1 tschetwerik, 2 t. = 1 payak,   -   = 1.438
  2 p. = 1 osmin, 2 o. = 1 tschetwerk,   -   -= 5.952
  1¼ tschetwerk = 1 kuhl, 16 tschetwerk = 1 last.
  Libau, Revel, Riga: 12 stof = 1 kulmet, 3 k. = 1 lof,
  24 l. = 1 tonne, 2 t. = 1 last.
SARDINIA I. — *Cagliari*, &c. : 4 imbuto = 1 carbula,
  4 c. = 1 starello, 3 s. = 1 restiere *or* rasiern,   = 4.166
SICILY I. — *Palermo, Messina*: 2 stari = 1 modello, 4
  m. = 1 tomolo, 4 t. = 1 bisaccia, 4 b. = 1 salma, = 7.81
  16 tomoli grosso = 1 salma grosso,   -   -   = 9.72
SPAIN. — *Alicant*: 2 medio = 1 celemin, 4 c. = 1 bar-
  cella, 12 b. = 1 cahiz, -   -   -   -= 6.992
  Barcelona: 4 picolin = 1 cortain, 12 c. = 1 quartera,
  2¼ q. = 1 carga,,1⅗ c. = 1 salma, -   -   = 8.191
  Cadiz (*Standard of* CASTILE) : 2 medio = 1 celemin, 12
  c. = 1 fanega, 12 f. = 1 cahiz, -   -   -= 19.189
  4 cahiz = 1 last,   -   -   -   -   = 76.759
  Corunna, Ferrol: 4 celemine = 1 ferrado, 3 f. = 1
  fanega, 12 fanega = 1 cahiz,   -   -   -= 19.189
  Gibraltar: Same as *Cadiz*.
  Malaga: Same as *Cadiz*.
  Santander: 144 celemin = 12 fanega = 1 cahiz -   = 24.984
  Valencia: 8 medio = 4 celemin = 1 barchilla, 12 bar-
  chille = 1 cahiz, -   -   -   -= 5.758
SWEDEN. — *Stockholm*, &c. : 2 stop = 1 kanna, 1¼ k.
  = 1 kappe, 4 kappe = 1 fjerding, 4 f. = 1 spann,
  2 spann = 1 tunna, -   -   -   -   = 4.158
  24 tunne = 1 last.
SWITZERLAND (legal, since 1823, for the Cantons of
  *Aarau, Basle, Berne, Freiburg, Lucerne, Solothurn,
  Vaud;* but not in general use) :
  10 emine = 1 gelt *or* quarteron, 10 g. = 1 sac =
  1$\frac{7}{20}$ *hectolitres* of France, -   -   -   -= 3.831
Special and local —
Basle: 2 bacher = 1 kopflein, 8 k. = 1 sester, -   = 0.97
  4 sester = 1 sack, 2 s. = 1 vierzel,   -   -= 7.756
Berne: 4 achterli *or* 2 immi = 1 massli, 2 m. = 1
  mass, 12 mass = 1 mut,   -   -   -   = 4.771

E

50 a FOREIGN DRY MEASURES REDUCED TO UNITED STATES.

Foreign.	U. States.
	Winchester bushels.
Geneva: 2 bichet = 1 coupe *or* sac,	= 2.203
Lausanne: 10 copet = 1 emine, 10 e. = 1 sac,	= 3.831
Neufchatel: 3 copet = 1 emine, 8 e. = 1 sac,	= 3.459
3 sacks = 1 muid.	
St. Gall: 4 massli = 1 vierling, 4 v. = 1 viertel, 4 viertel = 1 mutt, 2 m. = 1 malter,	= 4.688
Zurich: 16 massli *or* 4 vierling = 1 viertel, 4 v. = 1 mutt, 4 m. = 1 malter,	= 9.333
1 mass, for salt,	= 2.622
4 mass = 1 korb.	
TRIPOLI (N. Africa): 20 tiberi = 1 caffiso,	= 1.154
4 orbah = 1 tomen, 3½ t. = 1 nusfiah,	= 1.0157
3 n. = 1 ueba.	
TUNIS.— 12 zah *or* saha = 1 quiba, 16 q. = 1 caffiso = 18 wage,	= 14.954
TURKEY. — *Constantinople:* 4 kiloz = 1 fortin,	= 3.764
Latakia, Aleppo: 1 garave,	= 41.15
Smyrna: 4 kilo = 1 fortin,	= 5.824

West Indies.

In the Islands of *Antigua, the Bahamas, Barbadoes, Barbuda, Dominica, Grenada, Jamaica, Les Saints, Montserrat, Nevis, St. Kitts', St. Vincent, Tobago, Tortola,* the Dry Measures are the same as those of the United States.

In *Descade, Guadeloupe, Mariegalante, Martinique, St. Lucia:* 3 boisseau = 1 minot, 2 m. = 1 mine, 12 mines = 1 setier, 2 s. = 1 muid, - = 53.153

This being the system of France before 1812.

In *Bonaire, Curacoa, Saba, St. Eustatius, St. Martin:* Same as in Holland before 1820: See HOLLAND.

In *Santa Cruz, St. John, St. Thomas:* Same as DENMARK.

In *St. Bartholomew:* Same as SWEDEN.

In *Trinidad:* Same as CASTILE (*Spain*).

CUBA I. — Same as CASTILE, *generally.*
  Havana: 4 arrobas = 1 fanega, - = 3.114

HAYTI I. — *Aux Cayes, Cape Haytien, Port au Prince, Port Platte,* &c.: Same as France before 1812.
  Savanna, St. Domingo, &c.: Same as CASTILE.

PORTO RICO I. — Same as CASTILE (*Spain*).

CUSTOM HOUSE ALLOWANCES ON DUTIABLE GOODS.

Draft, or *Tret*, is an allowance of weight for supposed waste on articles paying duty by the pound. It is deducted from the *actual* gross weight of the article, and is established as follows:—

	Cwt.		Cwt.	Draft.
On	1 (112 lbs.)			1 lb.
Above	1 and under	. .	2,	. 2 "
On	2 " "	. . .	3,	. 3 "
"	3 " "	. . .	10,	. 4 "
"	10 " "	. . .	18,	. 7 "
"	18 " upwards,			9 "

Tare is the weight — actual or assumed by law — of the cask, sack, &c., in which the article paying duty is contained. It is deducted from the actual gross weight less the draft. The remainder is the *net* weight on which the duty is assessed, and the weight at which the heavy purchasers receive the goods.

Leakage is an allowance on the gauge of molasses, oils, wines, and all liquids in casks. It is established at 2 per cent., and is deducted from the actual gross gauge, less the real wants of the cask.

Breakage is an allowance of 10 per cent. on ale, beer and porter, in bottles, and 5 per cent. on all other liquors in bottles; or, if the importer prefer, the duties are assessed by actual count, he so electing at the time of making the entry. Common sized bottles are computed to contain 2⅜ gallons per dozen.

On bottles in which wine is imported there is assessed a duty of two dollars per gross, in addition to the duty on the wine.

The following articles, whether intended for sale or otherwise, are admitted into the United States, from foreign ports, free of duty; but nevertheless must pass through the Custom House in manner the same as goods on which a duty is assessed.

Animals imported for breed.
Antiquities.
Bulbs or bulbous roots.
Bullion, silver or gold.
Canary seed.
Cardamon seed.
Coins, gold, silver, or copper.
Copper sheathing, 14 by 48 inch, and from 14 oz. to 34 oz. per square foot.
Copper ore.
Cotton.

Cummin seed.
Fossils.
Gold dust.
Guano.
Gypsum, unground.
Oakum and old junk.
Oysters.
Platina, unmanufactured.
Silver, old, fit only for re-manufacturing.
Vanilla, plant of.

TABLE OF ESTABLISHED TARES.

(a=by custom; c=legal.)

Article	Per cent.	lbs. per pkg.
Almonds, in bags,	a 4	
Alum, casks,	a10	
Beef, jerked, drums,		70
" " hhds.,		112
Bristles, Archangel,	a14	
" Cronstadt,	a12	
Camphor, crude, tubs,	a35	
Candles, boxes,	c 8	
" chests, 160 lbs.,	a20	
Candy, sugar, baskets,	a 5	
" " boxes,	c10	
Cheese, hps. or baskets,	a10	
" boxes,	a20	
Chocolate, boxes,	c10	
Cinnamon, mats,	a10	
" chests,		16
Cloves, casks,	a12	
Cocoa, bags, (*actual* 2)	c 1	
" casks,	c10	
" zeroons,	a 8	
Coffee, E. I., grass bags,		2
" " bales,	c 3	
" " casks,	c12	
" W.I., bags,	c 2	
Copperas, casks,	a10	
Cordage, lines, bales,	c 3	
Cordage, mats,	a1¾	
Corks, bales, light,		5
" heavy,	a20	
Cotton, bales,	c 2	
" zeroons,	c 6	
Currants, casks,	a12	
Figs, boxes, 60 lbs.,	a10	
" ¼ " 30 "	a 6	
" " 15 "	a3½	
" drums,	a10	
" frails, 75 lbs.,	a 4	
Glue, Russia, boxes,	a15	
Indigo, bags or mats,	c 3	
" barrels,	c12	
" cases,	a15	

Article	Per cent.	lbs. per pkg.
Indigo, casks,	c15	
" zeroons,	c10	
Looking-glasses, Fr.,	a30	
Mace, kegs,	a33	
Nails, casks,	c 8	
Nutmegs, "	c12	
Ochre, French, casks,	a10	
Pepper, bales,	c 5	
" bags,	c 2	
" casks,	c12	
Pimento, bales,	a 5	
" bags,	c 3	
" casks,	c16	
Prunes, boxes,	a 8	
Raisins, Malaga, boxes,	a15	
" " casks,	a12	
" " jars,	a	18
" Smyrna, casks,	a12	
Salts, glauber, casks,	c 8	
Shot, casks,	c 3	
Soap, French, boxes,	a12	
" boxes, (*a* more)	c10	
Steel, bundles,	a 3	
" cases,	a 8	
Sugar, bags or mats,	c 5	
" boxes,	c15	
" casks,	c12	
" canisters,		40
" Java, willow baskets,		60
Tallow, casks,	a12	
" zeroons,	a 8	
Tea, caddies, } actual	about	6
" chests, } weight.	about	22
" ¼ "	about	14
Twine, bales,	c 3	
" casks,	c12	
Wool, Germany, bale,	a 3	
" S. Amer., bale,		15
" Smyrna, "		10

www.ingramcontent.com/pod-product-compliance
Lightning Source LLC
Chambersburg PA
CBHW031826230426
43669CB00009B/1235